SPACE

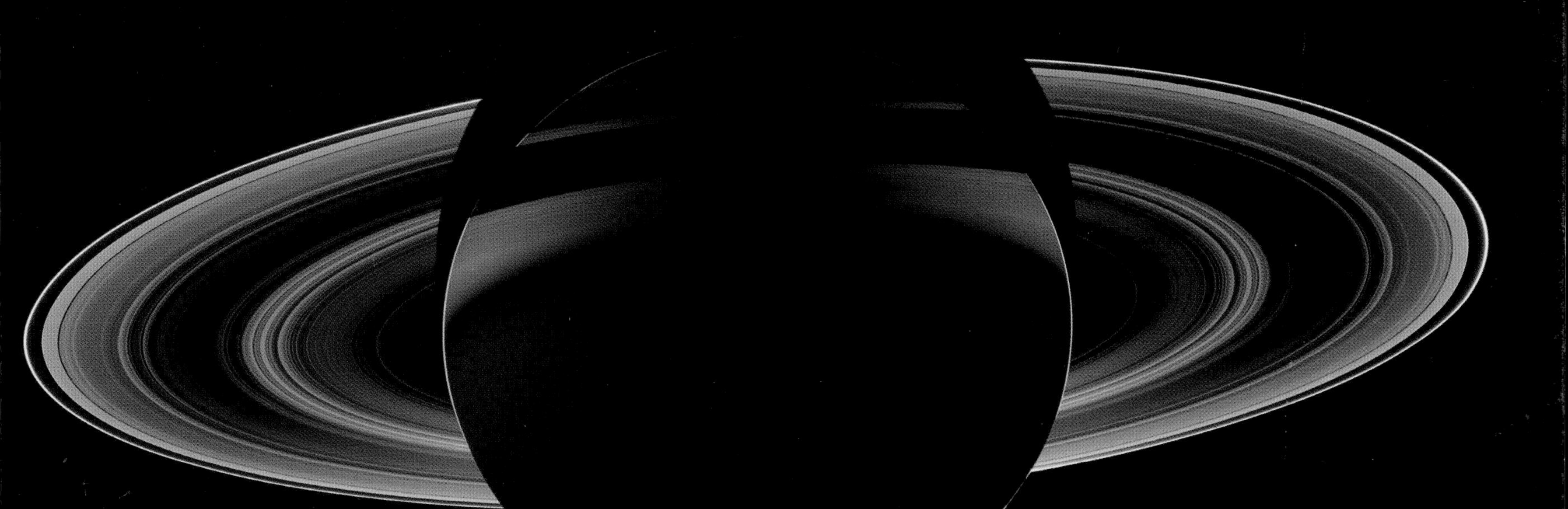

Sean Callery and
Miranda Smith

SCHOLASTIC

SCHOLASTIC

Art director: Bryn Walls

Managing editor: Miranda Smith

Consultant: Colin Stuart, MSc, Fellow of the Royal Astronomical Society

Contributors: Dr. Margaret Geller, Jefferson Hall, Professor Paul Hickson, Dr. Rosaly Lopes, Jerry L. Ross

ISBN 978-1-338-29196-4

10 9 8 7 6 5 4 3 2 1   18 19 20 21 22

Printed in China   38
First edition, September 2018

Detailed, all-sky picture of the infant Universe 13.7 million years ago

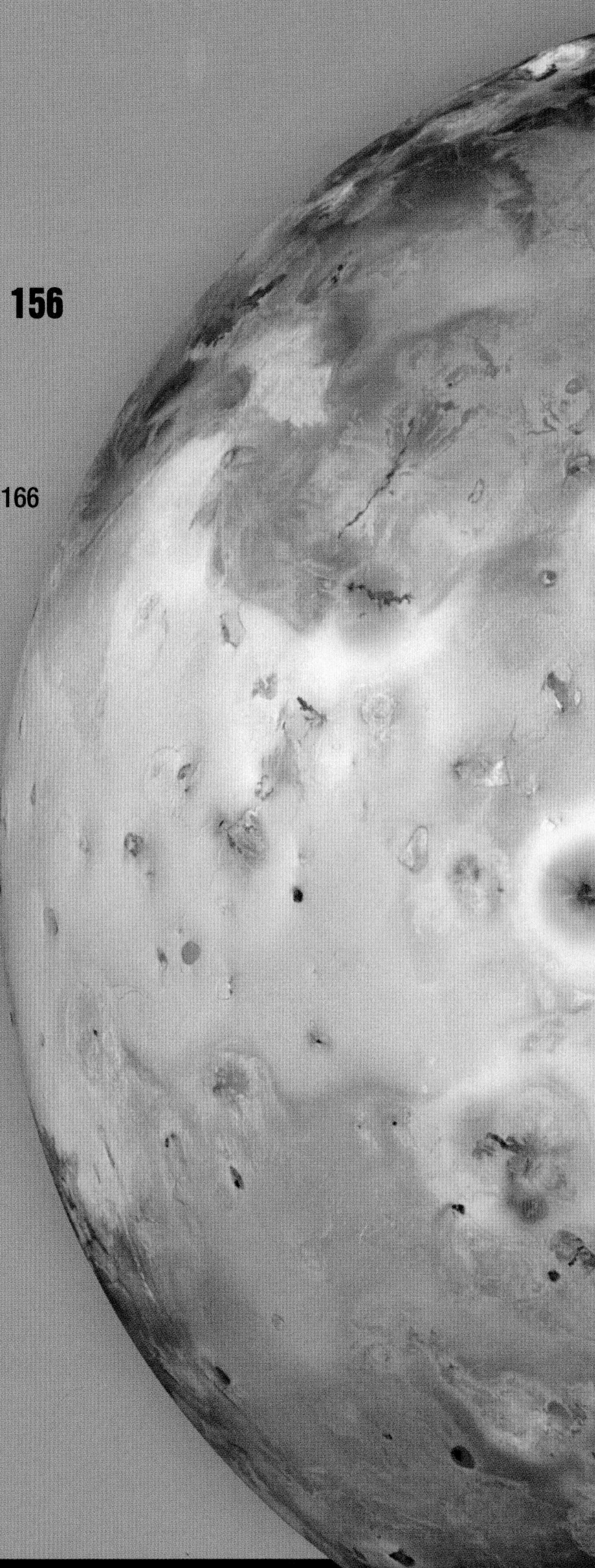

CONTENTS

FOREWORD

We are all on a journey through space as our spinning planet hurtles in a giant circle around the Sun. We are also on a journey of discovery about the Universe, and this book—your paper spaceship—is part of it.

We have discovered that out in the endless, silent cosmos sit countless stars, planets, moons, lumps of rock, and clouds of dust and gas. Everything, from the stars that dwarf our Sun to the tiniest speck of grit, is held in place and pulled around by the mysterious force of gravity.

It is a fascinating and bewildering journey across incredible distances. We have uncovered strange things such as red dwarves and black holes, and discovered that the volcanoes which shaped our planet are not unique. We have learned to explore through ideas, using what we know to develop theories about the rest. Books like this one show us how far we have come and point us toward what we seek to know. I hope you enjoy the ride!

Rosaly Lopes

Dr. Rosaly Lopes

Senior Research Scientist
NASA's Jet Propulsion Laboratory, Caltech

The beautiful barred spiral galaxy NGC 1300, in an image taken by the Hubble Space Telescope, lies about 70 million light-years away from Earth in the constellation Eridanus.

THE SOLAR SYSTEM

Earth is the third planet out from our Sun, with five more beyond us in a solar system that forms a tiny part of the Milky Way galaxy. A solar system is a star—in our case the Sun—and all the objects such as planets, moons, and asteroids, that orbit it. Our solar system is just one of many in infinite space.

DANGEROUS SCIENCE

The Earth goes around the Sun. Saying that sentence a few centuries ago would have led to you being laughed at (if you were lucky) or burned alive as punishment. Learning the truth about our solar system and understanding how it works has been a long, painful journey. The first civilizations noticed how the Sun crosses the sky each day and it was logical for them to assume that this great source of warmth and light travels around the Earth. It took many centuries (barely the blink of an eye in the life of the Universe) to realize that in fact the Earth and other objects in our solar system orbit the Sun.

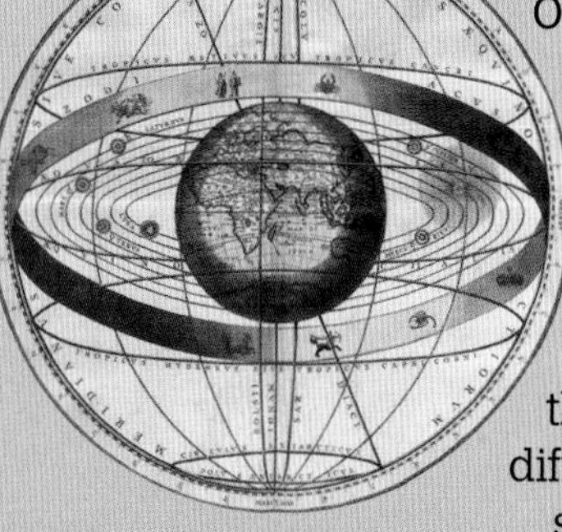

Early ideas
The Roman citizen and mathematician Ptolemy said that Earth was the center of the Universe. This model is of what came to be known as the Ptolemaic system.

PLANETS IDENTIFIED

Over thousands of years stargazers noticed that some objects in the sky were not stars, but something else. Mercury, Venus, Mars, Jupiter, and Saturn are all visible to the naked eye, and their positions change in a different way to those of the stars, so the ancient Greeks called them planets, meaning "wanderers." In the 16th century, astronomer Nicolaus Copernicus was the first to suggest the shattering idea that there were other planets in the solar system, and that they, together with Earth, orbited a star, the Sun.

THE SUN'S INFLUENCE

Centuries of study and thought have helped us understand why this is true. In the emptiness of space everything, from enormous lumps of rock or clouds of gas to tiny specks of dust, is pulled by a force called gravity. In our corner of the Universe, the Sun—the largest mass in our solar system and a million times bigger than Earth—has the strongest gravitational pull.

At the moment, we know that our solar system contains 1 star, 8 planets, 5 dwarf planets, 181 moons, nearly a million asteroids, and millions of comets. All are ruled by the Sun's gravity. The Earth and other planets were formed when gravity pulled rock and gas together from around the infant Sun.

Juno probe
This probe has been orbiting the planet Jupiter since 2016. Once every 53 days it is within only 2,600 miles (4,180 km) of Jupiter's cloud tops. It has sent back amazing new data showing how the vast winds flowing on the planet's surface extend deep into its interior, and that there are giant cyclones at the poles.

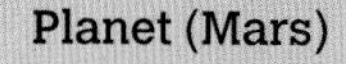

Planet (Mars)

Dwarf planet (Ceres)

Asteroid (Vesta)

Comet (67P/ Churyumov–Gerasimenko)

Objects in space
Everything in our solar sytem orbits, either directly or indirectly, a star, our Sun. Beyond Earth are comets, asteroids, and other rocky, frozen objects yet to be discovered in the Kuiper Belt and Oort Cloud.

Exploring space
From 1961 onward, we have been able to launch people into space. They have been able to look back at Earth, walk on the Moon, and work in laboratories on space stations to better inform us all about our world.

PLANET EARTH

Earth is a lump of rock weighing nearly 6 billion trillion tons, and spinning as it orbits 93 million miles (150 million km) round a star whose heat provides most of its energy. Earth rests in the perfect position for life—warmed but not roasted by the Sun, with a protective atmosphere holding oxygen, and plenty of water.

BIRTH OF A PLANET

About 4.6 billion years ago, gravity drew together a massive whirlpool of dust, metal, and rock that gained mass and developed gravity, forming a hot, dense core surrounded by molten rock. Earth was born. As the planet cooled, its surface hardened into a crust. After about a hundred million years, a lump of material roughly the size of Mars crashed into Earth, sending out chunks whose flight was halted by Earth's gravity. They became the Moon.

EARTH'S COMPOSITION

In a way, the interior of our planet is as much a mystery as the galaxies billions of light-years away. We know it is 3,958 miles (6,370 km) from Earth's surface to its core, but we cannot make the journey because we would boil in rock soup. It is only in the last hundred years that we have realized that the center is a solid, very hot inner core, sitting inside a spinning, liquid outer layer, made of metallic elements, that we now know

Carrying life
Ingredients for life could have arrived on our planet on comets or fragments of asteroids. Analysis of meteorites shows that some carry organic molecules.

generates an immense, invisible magnetic shield around our planet. Earth's crust is around 18 miles (30 km) thick under the continents, dropping to roughly 3 miles (5 km) deep under the oceans.

LIFE ON EARTH

Seen from a spacecraft or the Moon, Earth appears to be a blue planet. That is because about seven-tenths of it is covered in liquid water. No one knows for sure where Earth's water comes from. This water is essential for life, though life on Earth flourishes for a host of other reasons. These include the presence of oxygen in water and the atmosphere, and smaller amounts of carbon dioxide that moderate Earth's temperatures, as well as the process of photosynthesis, in which plants use light from the Sun to grow, and return oxygen to the atmosphere.

Life on Earth includes intelligent human life. We have learned to see better and farther, and to puzzle out things that we cannot see. We can now predict when galaxies will collide and we know that in about 4 billion years our Sun will burn out. We have made machines to blast us beyond Earth's gravity. And we wonder if we are alone. Ours is not the only galaxy. Ours is not the only solar system. Perhaps in our "greatest adventure," we will find other kinds of life, far across the Universe.

Atmosphere
This stops the planet from overheating by day and freezing at night.

Crust
The outermost solid shell is made of rigid, brittle rock.

Mantle
This thick layer of rock makes up 84% of the planet's total volume.

Core
This is a liquid outer core surrounding a solid inner core.

Earth's structure
Earth's layered structure is made up of the core, mantle, and crust. The crust and upper part of the mantle is broken into tectonic plates that move slowly over the mantle.

Earth from space
Astronauts aboard the International Space Station (ISS) took this image of southern Scandinavia just before midnight under a full moon. The wavering lines of green aurorae hover in the atmosphere above snow-covered land to the north, contrasting with the blackness of the Baltic Sea (lower right), and the twinkling lights that show human habitation. The concentrated clusters of light in the center are the capital cities of Oslo, Norway, and Copenhagen, Denmark.

STAR

WATCH

SEEING INTO SPACE

The sky beyond our lump of rock has obsessed humans for thousands of years. We have always made tools from lumps of metal and rock that fell to Earth. Star-gazers have drawn maps to show patterns of stars at night. Later technologies such as telescopes, observatories, and satellite-based lenses reveal detail, and provide data to explain some of the mysteries of space.

James Webb Space Telescope mirror on handling frame

JPSS-2 polar observation satellite

Tektite from meteorite strike

Vintage constellation card

Astronomer Nicolaus Copernicus

Radiowave image of Saturn

Spitzer Space Telescope

McDonald Observatory

Isaac Newton's reflecting telescope

Perseid meteor shower

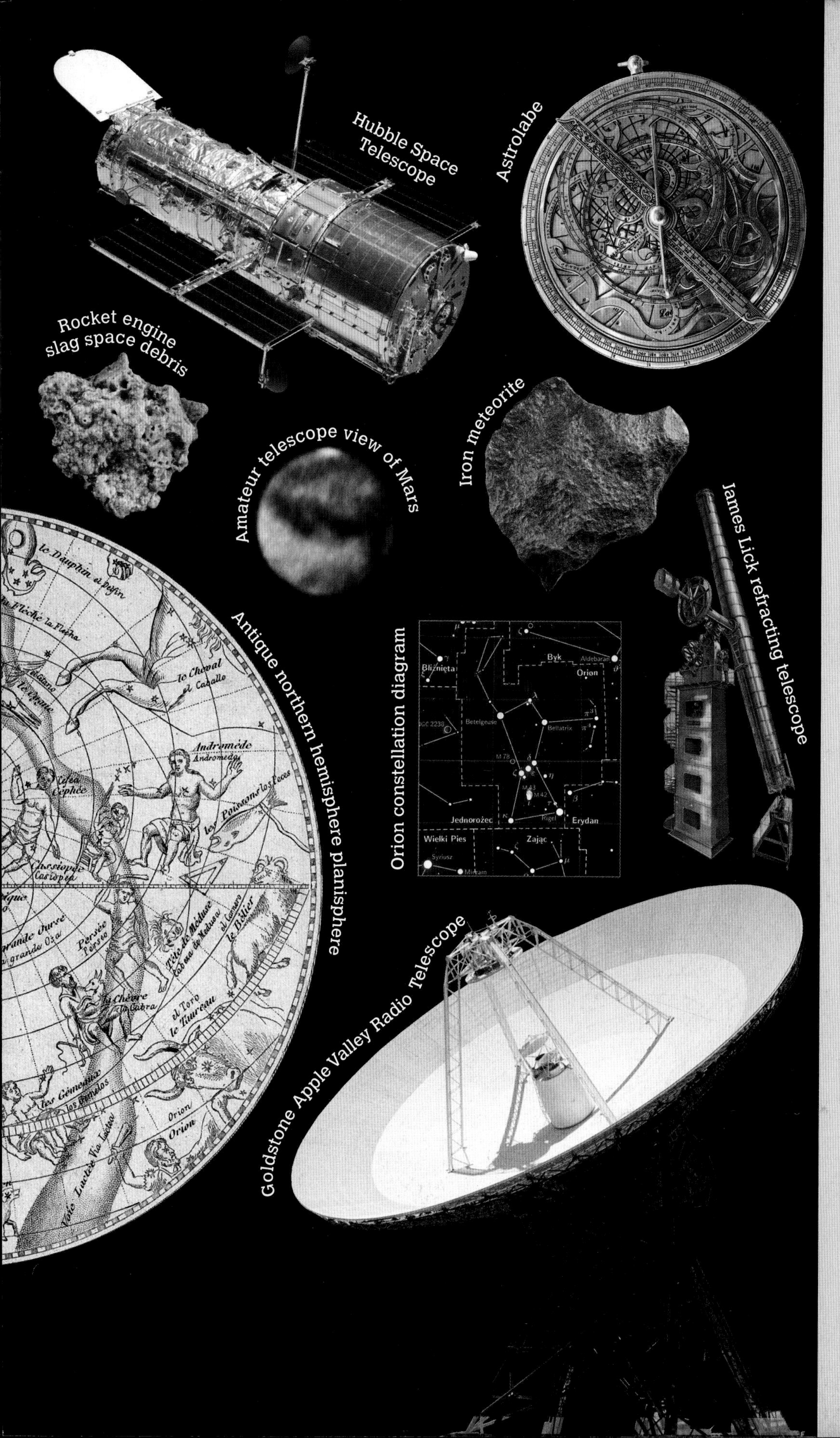

TOOLS FROM OUTER SPACE

Space has provided us with tools in lots of different ways. Meteorites are lumps of rock or metal zooming through space. Most burn up in the friction of Earth's atmosphere, but some survive and land. From earliest times, meteorites rich in iron were the raw material for metal tools and weapons. Later, people learned to heat rock and extract metal very efficiently.

SPACE-WATCHING

Tools became more sophisticated with time. The ancient Greeks forged astronomical instruments and used their math skills to calculate the size of the Earth. Their belief that the heavens revolve around our world was accepted as truth for 2,000 years. The invention of the telescope in the 17th century changed that, as glass lenses brought the realization that Earth and other planets orbit the Sun. More recently, astronomers have used ever-larger observatories, such as the Goldstone Apple Valley Radio Telescope, and have launched equipment including the Spitzer and Hubble space telescopes away from the blurring problems caused by Earth's weather and atmosphere. These tools allowed them to collect new forms of data such as X-rays, gamma rays, radio waves, microwaves, and ultraviolet and infrared radiation.

This image giving global sea temperature data was taken by an artificial satellite in orbit around Earth.

THE NIGHT SKY

MOONS, PLANETS, AND STARS

We can see parts of the story of the Universe from Earth with our eyes. The bright stars cluster together in galaxies and form patterns in the sky. The craters on the Moon remind us that objects flying through space can end their journeys with a crash. And the brightest dots are the five planets nearest to us, proof of other worlds out there.

Earth's Moon

The Moon (*see pp.70–73*) orbits Earth, so it is our planet's satellite, at its closest approach only 226,137 miles (363,932 km) away. At night, the Moon is the largest and brightest object in the sky, with its craters, mountains, and dark plains clearly visible on an unclouded night.

Draco

Ursa minor

Ursa major

Constellations

Since ancient times, people have seen patterns formed by the stars in the sky, calling them constellations (*see pp.30–31*). They gave them the names of animals, mythical beings, or familiar objects that were part of their everyday lives. Orion the Hunter, for example, was named by the ancient Greeks.

Aurorae

One of the most breathtaking events in the night sky is a display of dancing lights—aurorae—at the poles (left). When the Sun's activity is high, particles and radiation travel in solar winds through the solar system. When they reach Earth, they interact with its magnetic field, causing particles there to travel toward the poles and create the aurorae.

Comets
These small objects (*see p.117*) hurtling through space are made of ice and rock. They come from the distant edges of our solar system and, as they get closer to the Sun, they begin to break up, streaming out tails of dust and gas that we can see from Earth.

The **brightest star** in the **night sky** is **Sirius**, also known as the **Dog Star**.

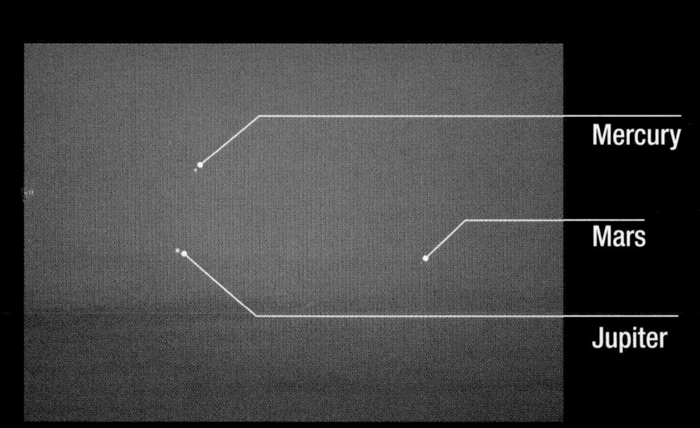

Planets
Planets (*see pp.54–55*) are close enough to shine steadily in the night sky—they do not twinkle like the more distant stars (*see p.175*). The five planets that can be seen with the naked eye are Mercury, Venus, Mars, Jupiter, and Saturn, and they can even be seen in light-polluted skies.

The Milky Way

Orion Nebula

Nebulae and galaxies
On a clear night, it is possible to see nebulae, the huge clouds of gas and dust in which stars are born (*see pp.170–171*). And there are at least 100 billion stars in our galaxy, the Milky Way (*see pp.164–165*), a beautiful river of stars that crosses the sky of both the northern and southern hemispheres.

Light show
On a dark, clear night you can see 3,000 stars with the naked eye. But our night skies are often filled with light pollution. The 24/7 cities take over the sky with artificial light, obscuring all but the brightest stars and planets.

Optical telescope

Binoculars

Seeing further
Your eyes can only gather a limited amount of light from faint objects in space, so it is possible see much farther with the aid of binoculars or an amateur telescope. The wider the aperature (the light-gathering opening) and the better the lens, the more you will see.

TELESCOPES

SPACE TECHNOLOGY

Telescopes gather in more light than reaches the eye, so we see more in greater detail. The early ones were polished glass lenses in a tube, then mirrors were used in ever-larger, more complex formations. Tubes and dishes became mountain-top observatories before we launched telescopes beyond Earth's light-bending atmosphere. By then, we were capturing not just light but radio waves and infrared rays.

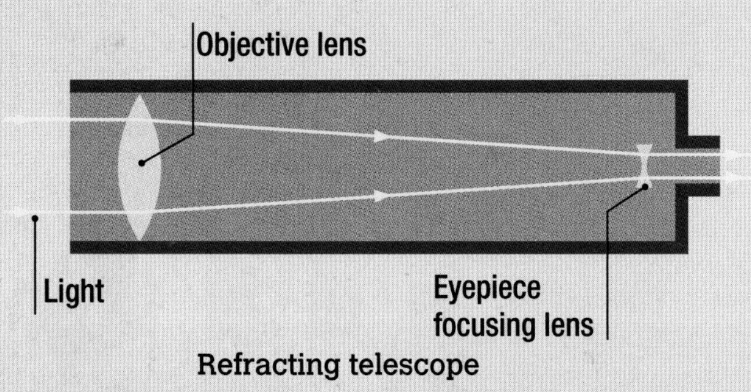

Refracting telescope

Secondary mirror
Primary mirror
Focusing lens

Reflecting telescope

"**Telescope**" is from the **Greek** ***tele*** meaning "**far**" and ***skopein*** meaning "**to look or see**."

First telescopes

The first telescopes were refracting telescopes that used lenses and were fairly simple to make. The two lenses refract or bend the light toward the eyepiece. The eyepiece then magnifies the image. Refracting telescopes are robust and lightweight but the images are always a little blurry.

Bigger and better

In reflecting telescopes, a convex mirror collects the light and a secondary mirror reflects it to a lens for a magnified image. These telescopes are more widely used by astronomers because the smooth surface of the mirrors reflects well, providing more detailed images. The bigger the telescope, the more it reveals.

Inspired invention

It is said that Hans Lippershey, a German-Dutch lens maker, was inspired to invent the refracting telescope when watching two children playing in his shop. They were able to make a weather-vane seem closer using two lenses. He filed a patent in 1608.

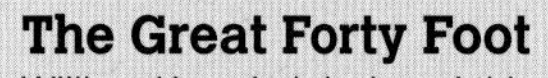

The Great Forty Foot

William Herschel designed this reflecting telescope in Slough, England. When it was finished in 1789, it was the biggest telescope in the world. William and his sister, Caroline, used it to find two new moons of Saturn.

Leviathan of Parsonstown telescope

In 1845, William Parsons, an Irish astronomer and engineer, completed the two-year building of a telescope that dwarfed all the others in the world at the time. With this reflecting telescope, he discovered 15 spiral galaxies, including M51, the Whirlpool galaxy. The mirror, like Newton's, was made of a mixture of copper and tin, which tarnished and had to be re-polished every six months.

Galileo

In 1609, Italian astronomer Galileo Galilei made the prototype of the refracting telescope used today. In it, the objective lens was convex and the eye lens was concave, whereas in a modern version there are two convex lenses.

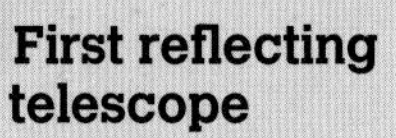

First reflecting telescope

In 1668, English astronomer Isaac Newton made a small but powerful reflecting telescope with a polished metal mirror. Having tried out various metals for the mirror, he settled on a mixture of copper and tin. The more light the mirror reflected, the better view there was of the sky.

Reber Radio Telescope

In the summer of 1937, US engineer and astronomer Grote Reber built the first radio telescope in his backyard. This pioneer in radio astronomy made a radio-frequency star map, completed in 1941 and extended in 1943. His data showed the brightness of the sky in radio wavelengths (*see p.22*).

Palomar 200-inch

Hale was not satisfied with a 100-inch telescope, so he secured funding and, in 1929, work began on the Palomar Observatory to house a 200-inch reflecting telescope which took 20 years to build. The telescope was dedicated in 1948, ten years after Hale's death. Among other discoveries, the telescope has helped astronomer Maarten Schmidt determine the nature of quasars, objects in deep space that are star-like.

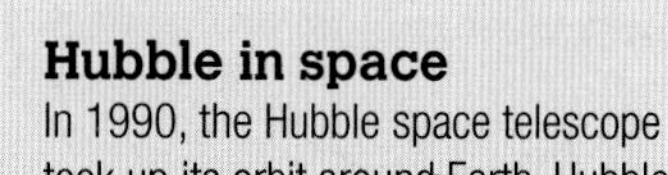

Hubble in space

In 1990, the Hubble space telescope took up its orbit around Earth. Hubble (*see pp.40–41*) has since beamed back hundreds of thousands of images that do not suffer from the distortion of Earth's atmosphere. It is a reflecting telescope and its two very smooth mirrors have precisely shaped reflecting surfaces.

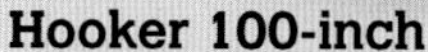

Hooker 100-inch

On July 1, 1917, the largest solid glass mirror ever made was transported to the Mount Wilson Observatory in California. It was to be fitted in a reflecting telescope designed by George Ellery Hale, and a local businessman, John D. Hooker, offered to pay for the production of the enormous 100-inch mirror. Edwin Hubble was using the telescope in the 1920s when he discovered that Andromeda was actually a galaxy in its own right.

Edwin Hubble uses the 100-inch Hooker telescope

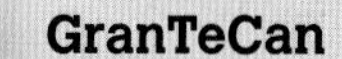

GranTeCan

Today, the world's largest optical telescope is the Gran Telescopio Canarias on La Palma in the Canary Islands. It is a reflecting telescope that has a mirror with a diameter of 34.1 ft (10.4 m), made up of 36 hexagonal pieces. These can be moved independently, and the shape of each can be changed. This helps to compensate for the turbulence in Earth's atmosphere.

SEEING FURTHER

TELESCOPES

The trouble with watching the stars from Earth is that you can only do it at night, and in good weather. But space-based telescopes can scan the darkness of space 24/7, in all weathers, away from the distorting effects of the Earth's atmosphere. They are also ideally placed to capture X-rays, infrared, ultraviolet, and radio waves. They radio their data back to Earth and allow us to build up the most detailed images so far of parts of the cosmos.

Different views of the Andromeda galaxy
The image below shows the Andromeda galaxy as it is seen by different telescopes on land and in space, able to "see" different electromagnetic wavelengths (*see below left*). The waves are able to transmit energy and travel through the vacuum of space.

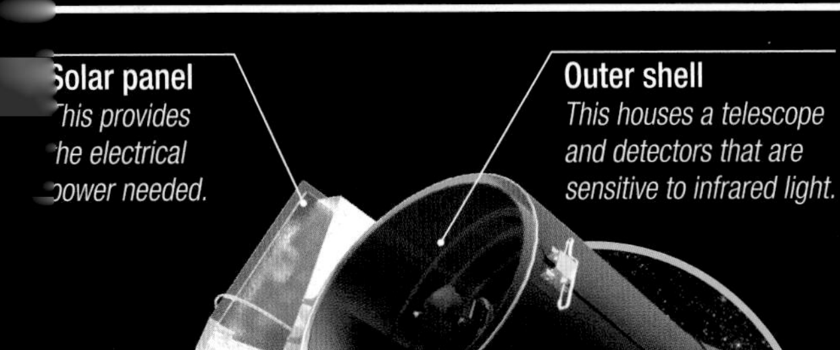

Solar panel
This provides the electrical power needed.

Outer shell
This houses a telescope and detectors that are sensitive to infrared light.

Spacecraft bus
This octagonal structure supplies the science instruments with power from the solar panels, and collects data to transmit.

Infrared—Spitzer
The Spitzer Space Telescope picks up the infrared light from space dust disks around distant stars. This allows astronomers to measure the temperature of the space dust by its color—the cooler dust is redder. The scientists then use the information to determine the structure and age of the disk.

Telescopes allow us to **"see"** parts of the **light spectrum** beyond **visible light**.

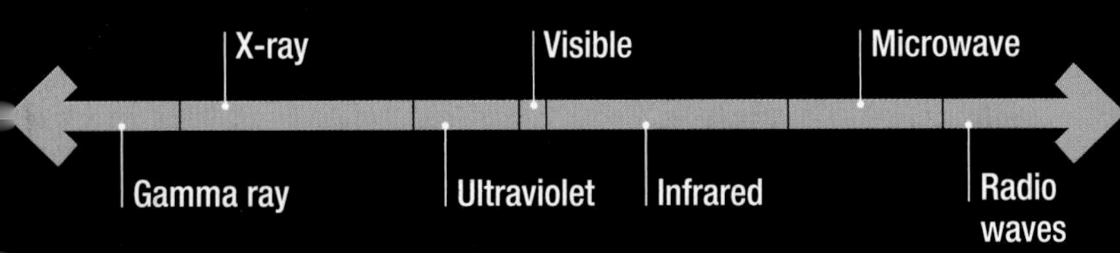

The wavelength spectrum
The energy produced by the Sun travels to Earth via electromagnetic waves that are different sizes. The shorter waves—these have the most energy—are gamma, x-ray, and ultraviolet; medium-sized waves are visible light; and the longest are infrared, microwave, and radio

Visible light—optical telescopes
Visible light astronomy can describe what is seen with the naked eye as well as what can be seen with Earth-bound optical telescopes. Sometimes on a clear night it is possible for amateur astronomers to obtain a great view of objects in space. However, far too often, turbulence in Earth's atmosphere obscures the view

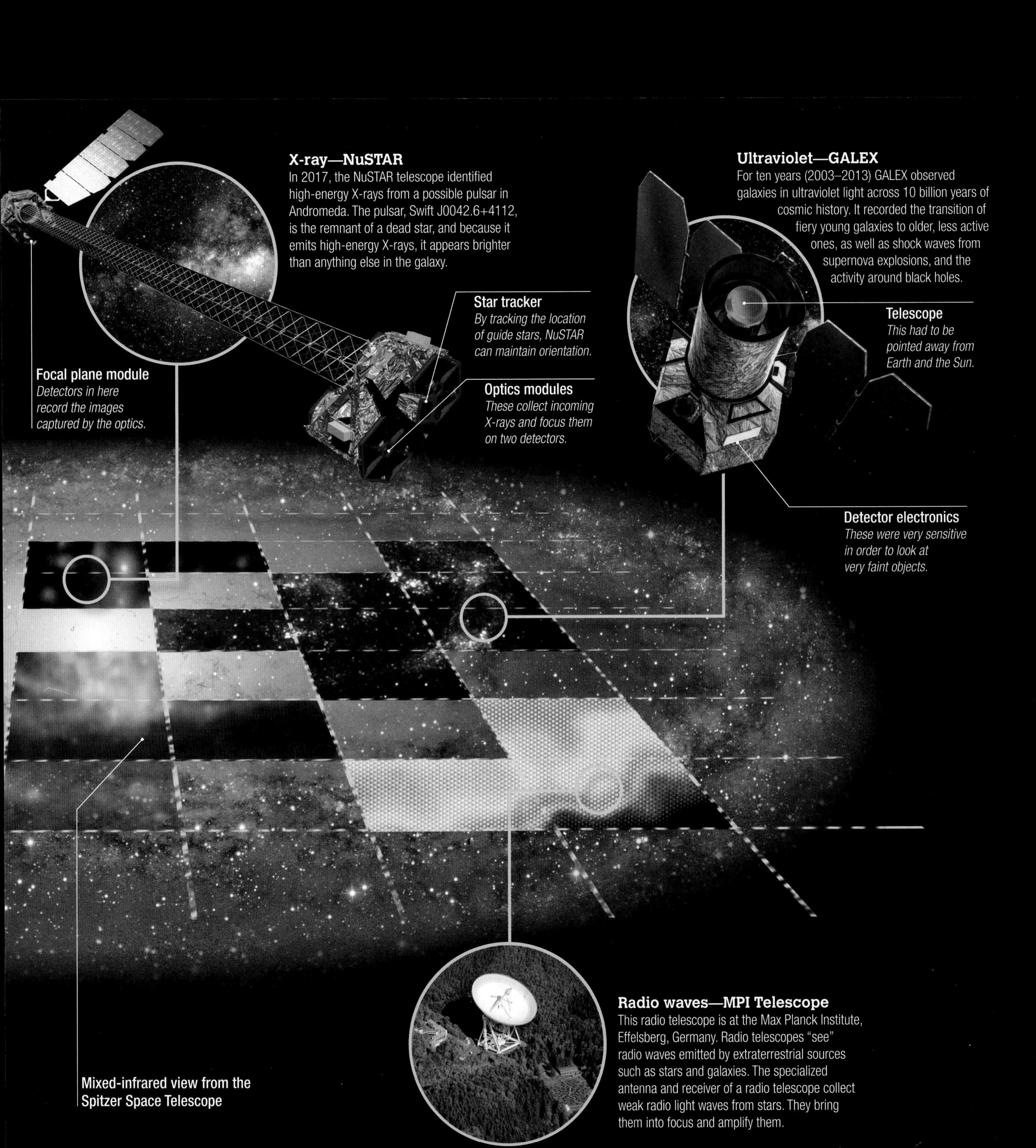

X-ray—NuSTAR
In 2017, the NuSTAR telescope identified high-energy X-rays from a possible pulsar in Andromeda. The pulsar, Swift J0042.6+4112, is the remnant of a dead star, and because it emits high-energy X-rays, it appears brighter than anything else in the galaxy.
Star tracker
By tracking the location of guide stars, NuSTAR can maintain orientation.
Focal plane module
Detectors in here record the images captured by the optics.
Optics modules
These collect incoming X-rays and focus them on two detectors.
Ultraviolet—GALEX
For ten years (2003–2013) GALEX observed galaxies in ultraviolet light across 10 billion years of cosmic history. It recorded the transition of fiery young galaxies to older, less active ones, as well as shock waves from supernova explosions, and the activity around black holes.
Telescope
This had to be pointed away from Earth and the Sun.
Detector electronics
These were very sensitive in order to look at very faint objects.
Radio waves—MPI Telescope
This radio telescope is at the Max Planck Institute, Effelsberg, Germany. Radio telescopes "see" radio waves emitted by extraterrestrial sources such as stars and galaxies. The specialized antenna and receiver of a radio telescope collect weak radio light waves from stars. They bring them into focus and amplify them.
Mixed-infrared view from the Spitzer Space Telescope

Laser sights

In the early morning light, the twin Keck telescopes near the summit of a dormant volcano, Mauna Kea, in Hawaii, beam lasers into the atmosphere. A technology called adaptive optics corrects for Earth's often turbulent atmospheric conditions, so that scientists can take sharper pictures of objects in space that were previously too blurry to be seen clearly. The adaptive optics measures and then corrects the disturbance with a deformable mirror that changes shape 2,000 times per second, providing an image clarity previously not possible.

NORTHERN SKIES

CONSTELLATIONS

European and Asian astronomers mapped the night sky above the northern half of Earth long ago, with its recognizeable patterns or constellations. They are grouped around the north celestial pole, a point in the sky about which all the stars in this hemisphere rotate. Polaris, or the North Star is very near this point.

Like the **Sun**, **constellations** travel from **east** to **west** across the **sky**.

M31, the Andromeda galaxy, in the constellation of Andromeda

1° 5° 10° 15° 25°

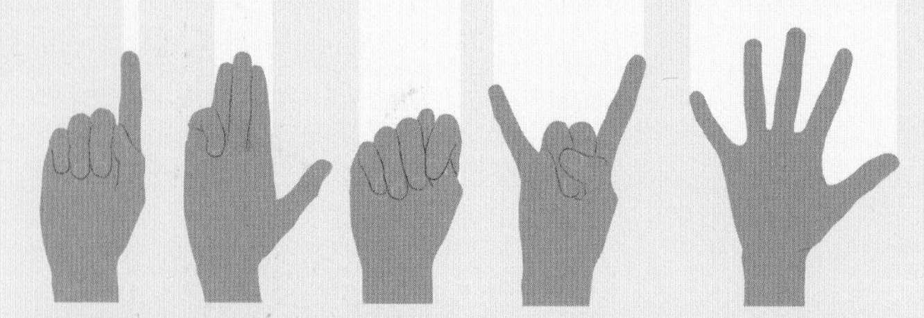

Measuring angles in the sky

The apparent distance between objects in the sky is measured in degrees, and you can measure the sky to find out approximate distances using your hands. For example, in the image below, the star on the left and the North Star (right) are three fists, or 30 degrees, apart in the night sky.

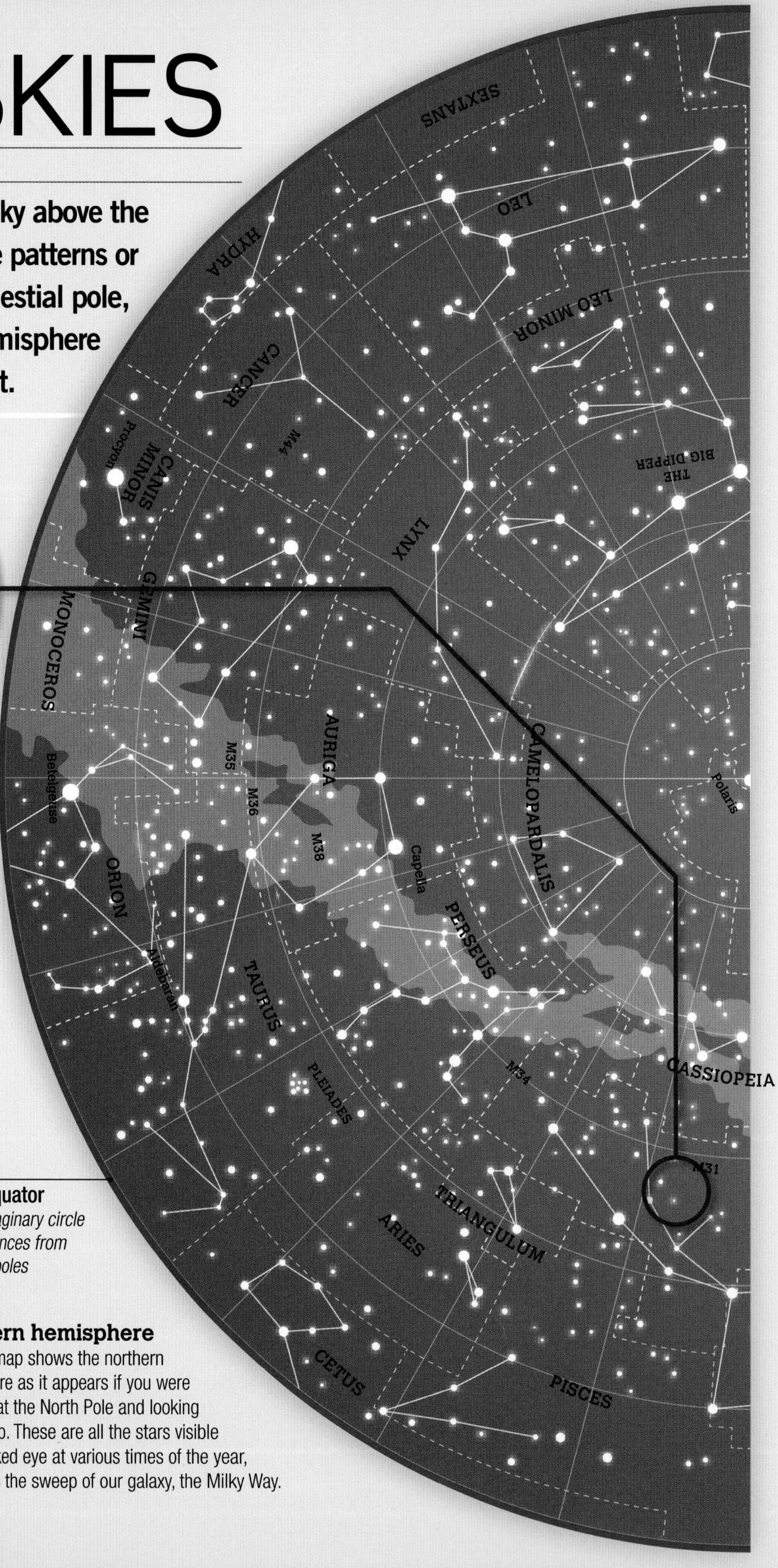

Celestial equator

This is an imaginary circle at equal distances from the celestial poles

Finding the North Star

Polaris, the North Star, has been used by sailors to navigate the seas for thousands of years. You can find it if you draw an imaginary straight line from part of the Big Dipper.

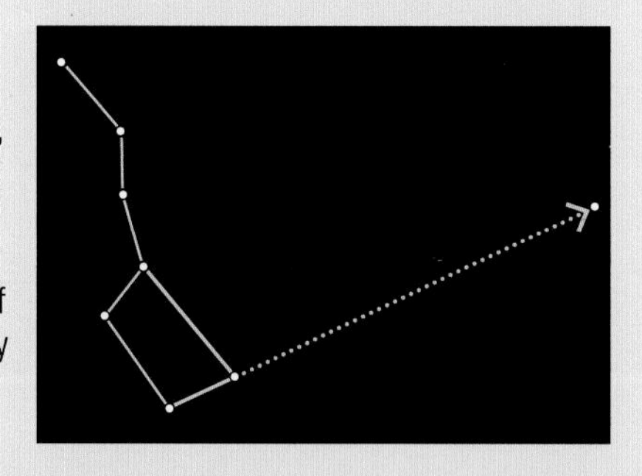

Northern hemisphere

This sky map shows the northern hemisphere as it appears if you were standing at the North Pole and looking straight up. These are all the stars visible to the naked eye at various times of the year, as well as the sweep of our galaxy, the Milky Way.

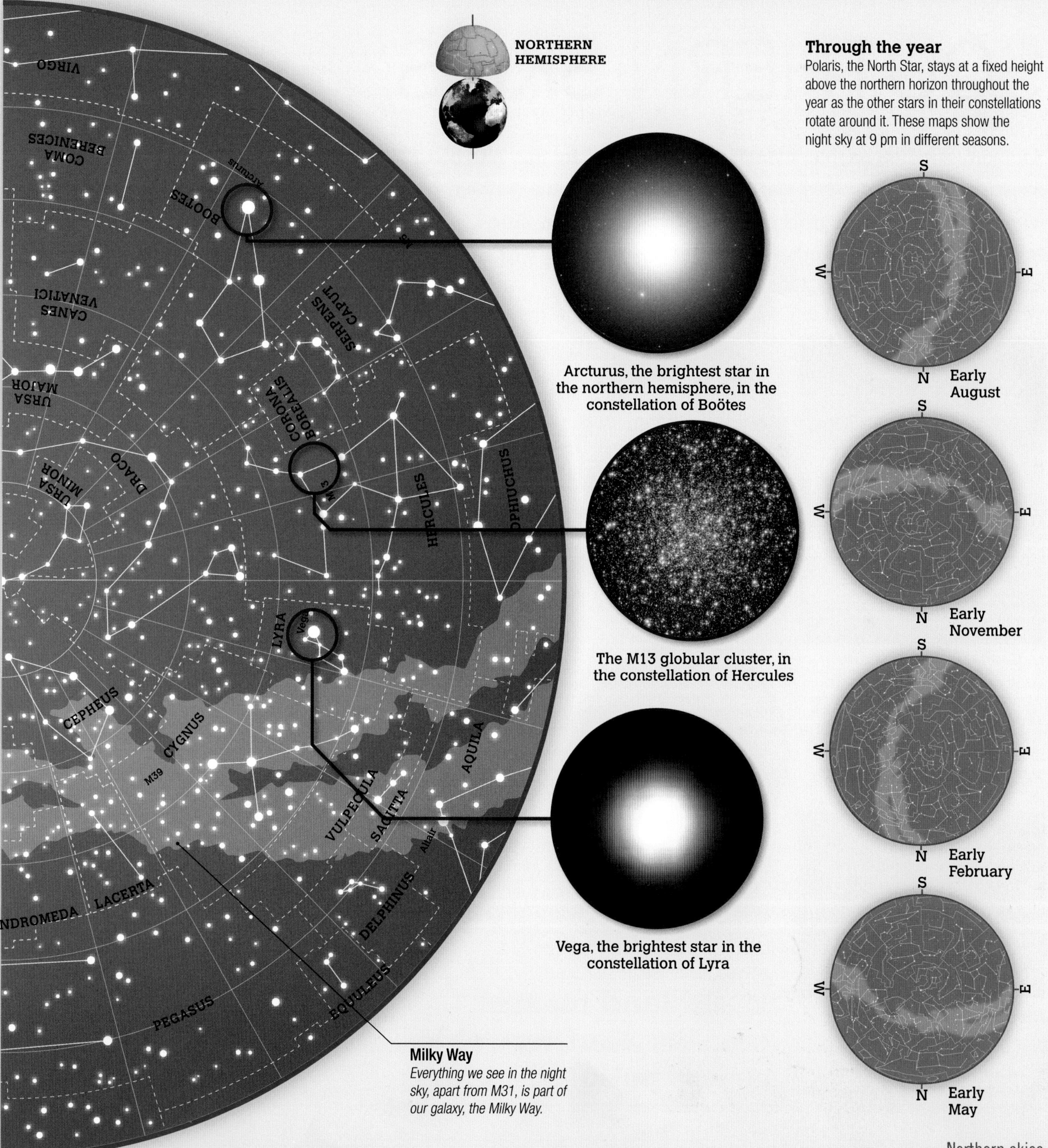

Through the year

Polaris, the North Star, stays at a fixed height above the northern horizon throughout the year as the other stars in their constellations rotate around it. These maps show the night sky at 9 pm in different seasons.

Arcturus, the brightest star in the northern hemisphere, in the constellation of Boötes

The M13 globular cluster, in the constellation of Hercules

Vega, the brightest star in the constellation of Lyra

Milky Way

Everything we see in the night sky, apart from M31, is part of our galaxy, the Milky Way.

SOUTHERN SKIES

CONSTELLATIONS

Southern skies are brighter than the northern skies, for they also boast the strange glow of the two Magellanic clouds. Where the northern hemisphere has Polaris as its guiding star, the south has Crux, or the Southern Cross, a tiny, five-star constellation that helps people navigate because it points to the south.

Galaxy's core
The Milky Way forms a great circle in the sky, so you can see equal amounts from the same place on Earth in either hemisphere. However, the center of the Milky Way is in the constellation of Sagittarius, so if you live in the southern hemisphere, it is easier to see.

Antares, a red star in the constellation of Scorpius

Center of the globular cluster M22 in the constellation of Sagittarius

Clear, desert air
The arid Atacama Desert in Chile is one of the best places for stargazers. The dry desert atmosphere means that there are more than 200 cloudless nights a year, providing a superbly clear view of the southern skies.

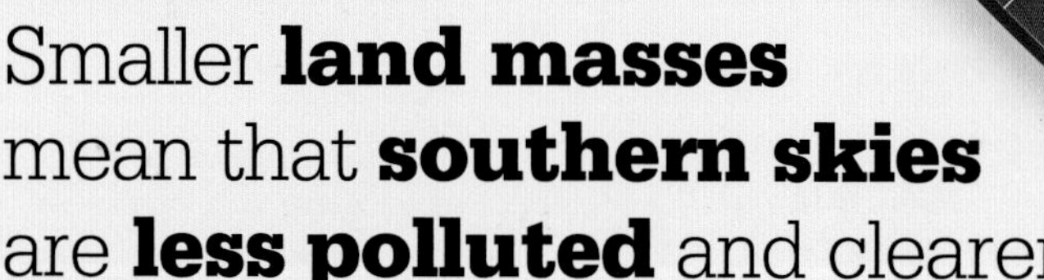

Smaller **land masses** mean that **southern skies** are **less polluted** and clearer.

Through the year

The south celestial pole (an imaginary point above Earth's south pole) stays at a fixed height above the southern horizon throughout the year as the other stars rotate around it. These maps show the night sky at 9 pm in different seasons.

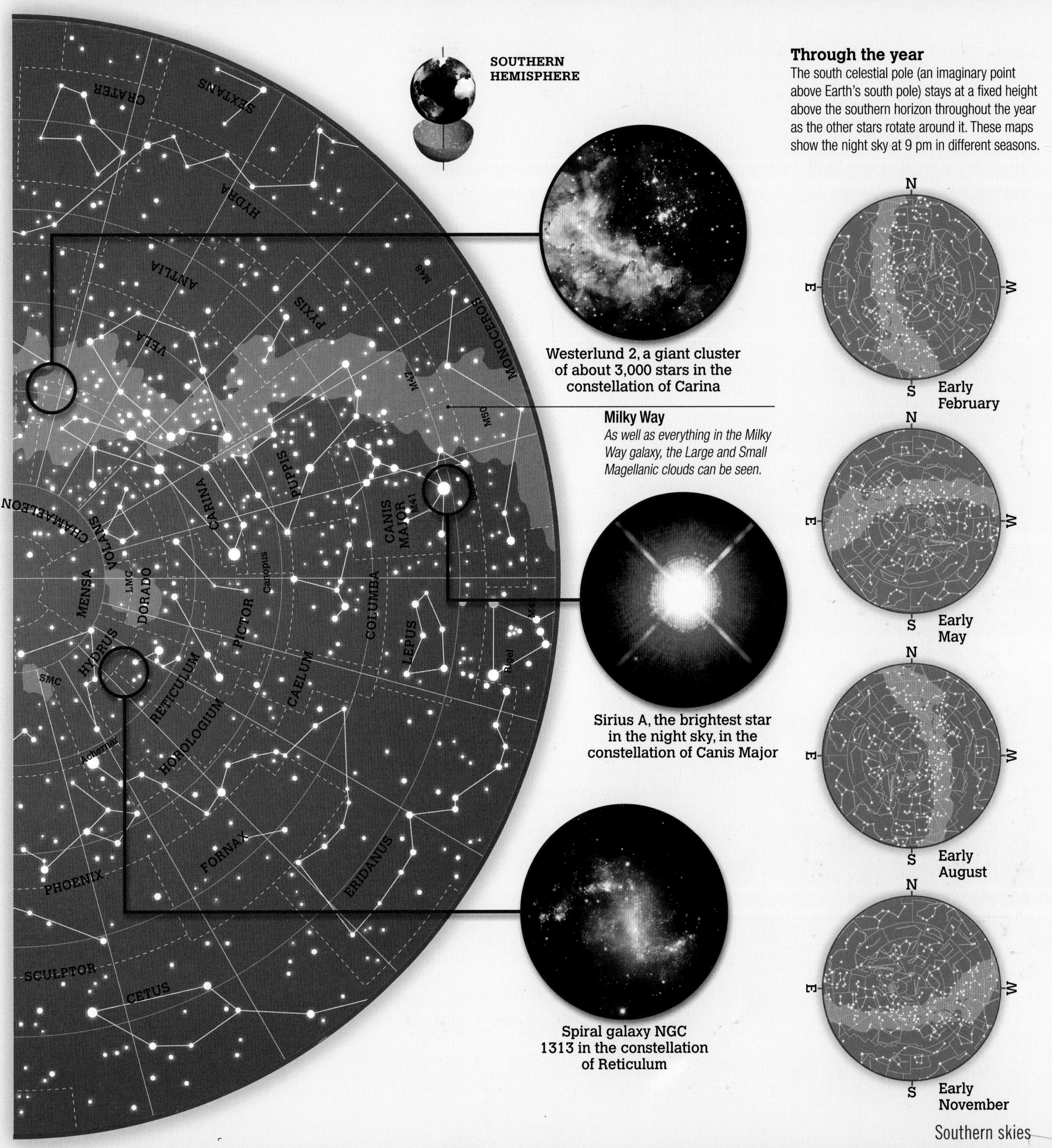

Westerlund 2, a giant cluster of about 3,000 stars in the constellation of Carina

Milky Way
As well as everything in the Milky Way galaxy, the Large and Small Magellanic clouds can be seen.

Sirius A, the brightest star in the night sky, in the constellation of Canis Major

Spiral galaxy NGC 1313 in the constellation of Reticulum

CONSTELLATIONS

STAR PATTERNS

Seen from Earth, some groups of stars make patterns that we call constellations. Ancient civilizations found they could use the shapes to keep track of the calendar and know the best times to plant and harvest crops. When, later, explorers sailed the oceans, they used the changing positions of the constellations to locate themselves and work out which direction to take. In 1922, 88 constellations were officially listed by astronomers, although more than half had been named by the ancient Greeks.

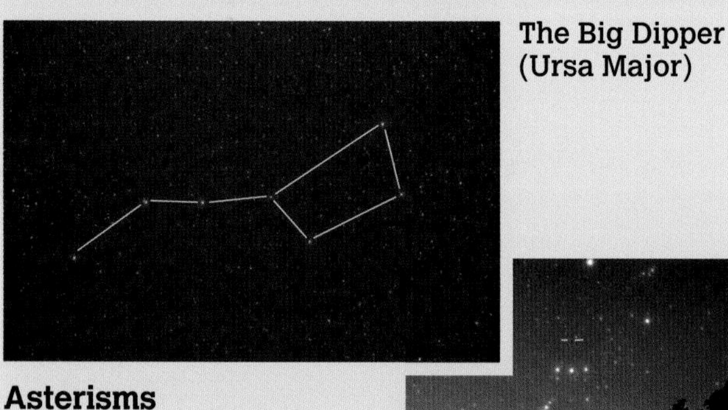

The Big Dipper (Ursa Major)

Orion's Belt (Orion)

Asterisms

Asterisms are distinctive shapes in their own right, but also part of larger constellations. For example, the popular Big Dipper is part of the constellation of Ursa Major, and three bright stars form Orion's Belt, part of the constellation of Orion.

Schedar

Cassiopeia northern skies

Named for a beautiful but vain Ethiopian queen in Greek mythology, this is noted for its asterism of five stars that form a giant M or W. Its brightest star, Schedar, is 40 times larger in diameter than the Sun.

Leo (the Lion) northern skies

This constellation, resembling a crouching lion, is one of the easiest to recognize in the night sky. Its brightest star is Regulus, where surface temperatures are more than twice those of the Sun. Leo was first cataloged by the Greek astronomer Ptolemy in the 2nd century CE.

Regulus

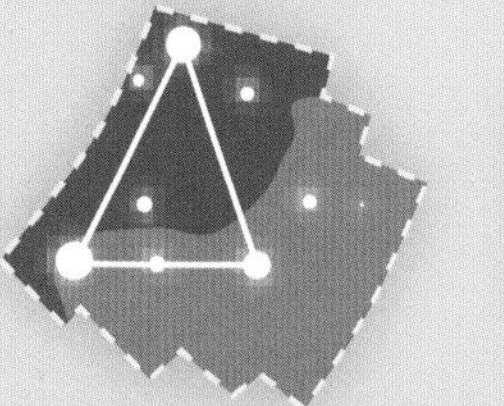

Triangulum australe

Crux

Small southerners

Two distinctive constellations are easy to spot in southern skies. Triangulum australe is named for its three-star triangle. The cross-shaped Crux is also known as the Southern Cross.

Alphard

Celestial equator

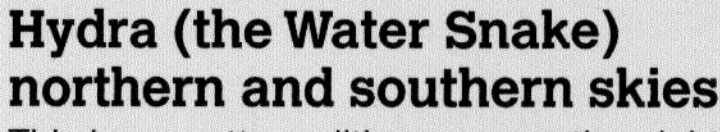

Hydra (the Water Snake) northern and southern skies

This huge pattern slithers across the night sky of the southern hemisphere, the largest of all the constellations. In the center Alphard, an orange giant, is its brightest star. It is named for the snake that the legendary hero Hercules had to slay as one of his 12 labors.

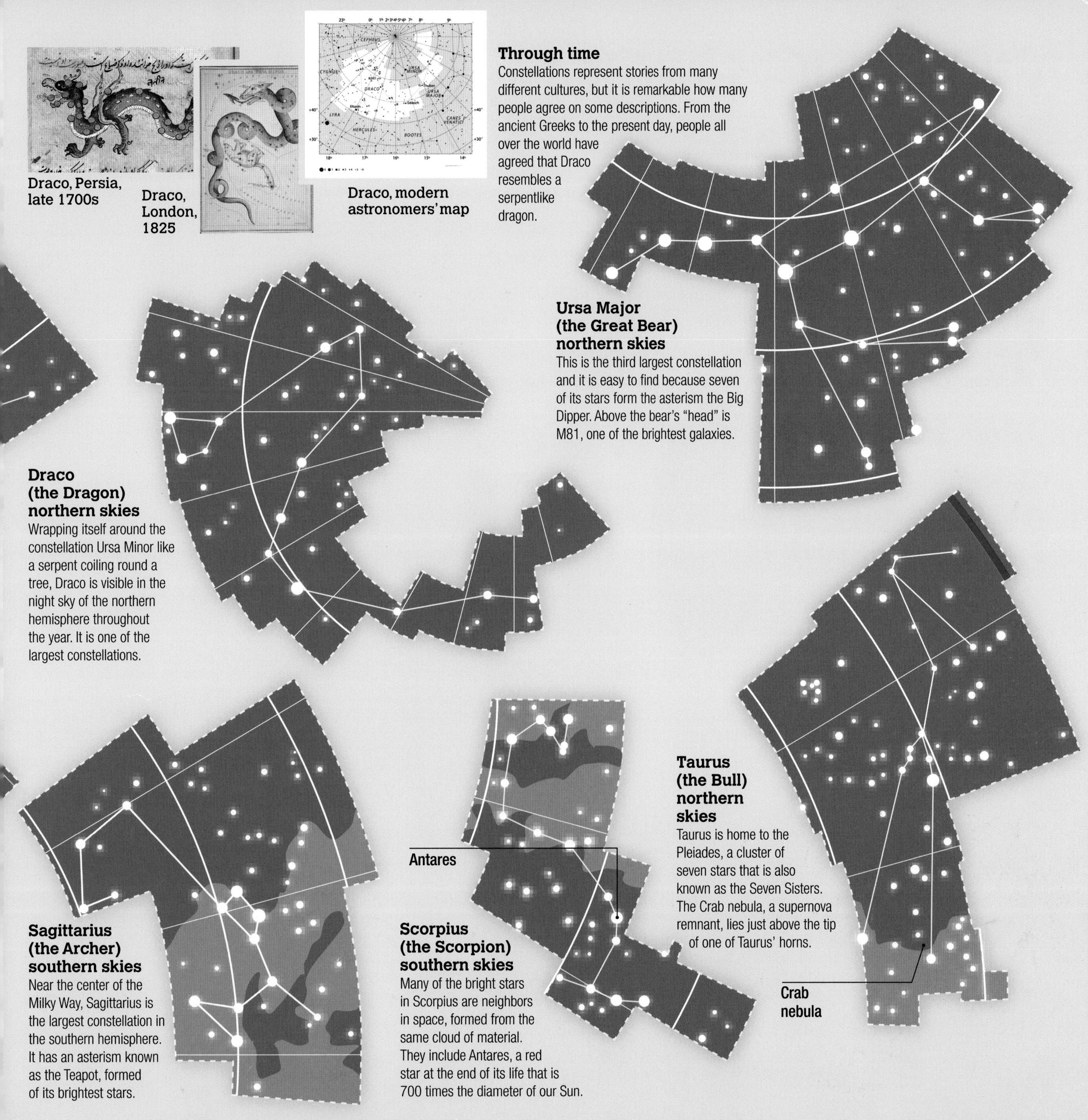

Draco, Persia, late 1700s

Draco, London, 1825

Draco, modern astronomers' map

Through time
Constellations represent stories from many different cultures, but it is remarkable how many people agree on some descriptions. From the ancient Greeks to the present day, people all over the world have agreed that Draco resembles a serpentlike dragon.

Ursa Major (the Great Bear) northern skies
This is the third largest constellation and it is easy to find because seven of its stars form the asterism the Big Dipper. Above the bear's "head" is M81, one of the brightest galaxies.

Draco (the Dragon) northern skies
Wrapping itself around the constellation Ursa Minor like a serpent coiling round a tree, Draco is visible in the night sky of the northern hemisphere throughout the year. It is one of the largest constellations.

Taurus (the Bull) northern skies
Taurus is home to the Pleiades, a cluster of seven stars that is also known as the Seven Sisters. The Crab nebula, a supernova remnant, lies just above the tip of one of Taurus' horns.

Sagittarius (the Archer) southern skies
Near the center of the Milky Way, Sagittarius is the largest constellation in the southern hemisphere. It has an asterism known as the Teapot, formed of its brightest stars.

Scorpius (the Scorpion) southern skies
Many of the bright stars in Scorpius are neighbors in space, formed from the same cloud of material. They include Antares, a red star at the end of its life that is 700 times the diameter of our Sun.

COMING FROM SPACE

SPACE OBJECTS

Some bright lights or flashes that you see in the sky are not planets or stars. They are comets, asteroids, or meteoroids that have hurtled through space for millions of years, and Earth is hit by some of them. Longer-lasting streaks with a tail are most likely comets (*see pp.116–117*). Scientists reckon that about 48.5 tons of rock and dust from these bodies strikes our planet every day. Most of it burns up when it hits our atmosphere and we say we have seen "shooting stars." The larger pieces that reach Earth's surface are called meteorites.

TYPES OF METEORITE

Metal: iron meteorite section showing long nickel-iron crystals

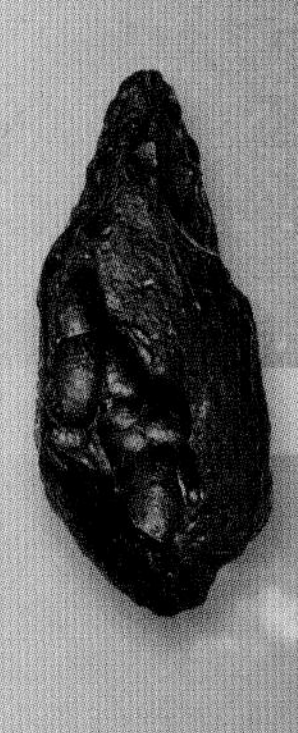

Flying space debris
Meteorites come from inside our solar system, and most are fragments of asteroids. They are either metallic, stony, or a mixture of both, depending on whether they come from the core or outside of an asteroid.

Stone: meteorite from Mars

Metal and stone: iron and obsidian meteorite

Flaming meteor
A large meteoroid can be traveling at tens of thousands of miles an hour when it hits Earth's atmosphere. It compresses the air in its path, heating it at temperatures of up to 3,000°F (1,650°C). These flaming streaks in the sky are called meteors.

Big impacts
Sixty-five million years ago an enormous asteroid slammed into Earth with an explosive impact estimated to be more than that of 100 trillion tons of TNT. The resulting darkness caused Earth's temperatures to plunge, and plants and smaller animals to die, with the result that the dinosaurs became extinct. This image is of the Meteor Crater in Winslow, Arizona. It is 570 ft (174 m) deep and 4,100 ft (1,250 m) across, and was caused by a meteor that was 130 ft (40 m) wide and traveling at 26,800 mph (43,130 km/h) reaching the surface some 50,000 years ago.

Meteor showers

At certain times in the year, Earth passes through space dust, and this causes beautiful displays of meteor showers, also known as "shooting stars." Most of these showers are caused by the icy debris that streams out behind a passing comet. This is a Leonids meteor shower, caused by debris from the comet Tempel-Tuttle.

Slag from rocket engine

Parts of a spacecraft

Beware flying rubbish

Not all objects hitting Earth come from outer space. There is plenty of manufactured debris hitting our atmosphere as well. Slag waste material from burning rocket fuel and even metal pieces from the rockets themselves sometimes make it to the ground.

Human waste in space

Some "shooting stars" are actually human waste matter being ejected from the ISS and hitting the atmosphere. Only solid matter is treated like this. Urine is recycled on board as water is too valuable to waste.

Comet 67P/Churyumov–Gerasimenko

Regular visitors

Many comets pass by Earth as they orbit the Sun, and some of their dust hits Earth's atmosphere. There may be many years between visits—Comet Lovejoy orbits every 8,000 years.

Comet Lovejoy

Most **space rocks** hit **our planet** at around **33,500 mph** (54,000 km/h).

Dangerous work
Astronomers attained a high rank in Chinese courts from early times. In ancient China, it was believed that events in the sky warned of events on Earth. Here, in this 17th-century tapestry, astronomers advise the emperor (standing with his hand raised), using a telescope and an armillary sphere, a model of objects in the sky (top right).

STARGAZERS

These extraordinary astronomers through the ages have made their mark with amazing discoveries that truly expanded our knowledge of the Universe and radically altered our view of our own planet. When, in 1543, Copernicus said that Earth was not the center of the Universe, this triggered a revolution not only in astronomy, but in religion, science, and society as people struggled to adapt to this new world view.

Early representations

The Nebra sky disk, found in a mine in eastern Germany, is 3,600 years old and the oldest depiction of the cosmos known. The Aztec Sun Stone, carved 3,000 years later, in 1479, emphasizes the importance of the Sun to the Aztec civilization.

Pleiades

Sun or full Moon

Crescent Moon

Bronze Age Nebra sky disk

Aztec Sun Stone

Aristotle

This Greek philosopher (384–322 BCE) helped to prove Pythagoras' theory that Earth is round. Aristotle noted that ships sailing over the horizon disappeared hull first, and that Earth cast a round shadow on the Moon during an eclipse. But he still believed that Earth was the center of the Universe.

Marble bust of Aristotle

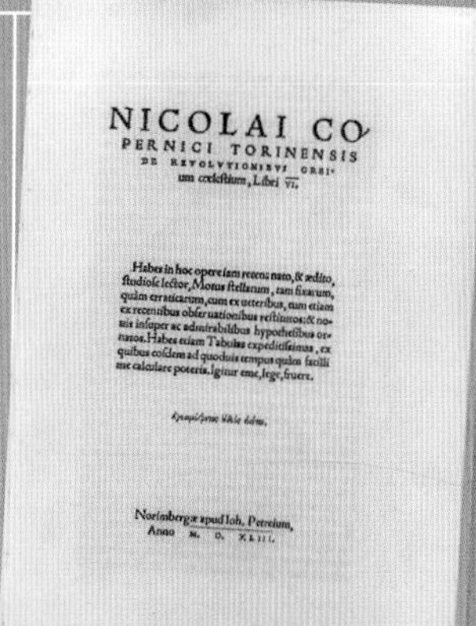

NICOLAI CO
PERNICI TORINENSIS
DE REVOLVTIONIBVS ORBI-
um cœlestium, Libri VI.

Habes in hoc opere iam recens nato, & ædito, studiose lector, Motus stellarum, tam fixarum, quàm erraticarum, cum ex ueteribus, tum etiam ex recentibus obseruationibus restitutos: & nouis insuper ac admirabilibus hypothesibus ornatos. Habes etiam Tabulas expeditissimas, ex quibus eosdem ad quoduis tempus quàm facillime calculare poteris. Igitur eme, lege, fruere.

Norimbergæ apud Ioh. Petreium,
Anno M. D. XLIII.

Copernicus

Nicolaus Copernicus was a Polish astronomer who argued that Earth rotates daily on its axis and yearly round the Sun. He placed the Sun at the center of the Universe, controversially challenging the theory that Earth held that position. His major work (left) was published in early 1543, the year he died.

Title page of Copernicus' "On the Revolutions of the Celestial Spheres"

Galileo stood trial in **1633** because he said **Earth** orbited the **Sun**.

Galileo Galilei

Galileo (1564–1642) invented his own telescope, making discoveries that revolutionized the astronomy of the period. He paved the way for people to accept Nicolaus Copernicus' theory that the Sun, not Earth, is at the center of the Universe.

Isaac Newton

This English physicist (1642–1727) is said to have discovered gravity when, at the age of 23, he watched an apple falling from a tree. From this, he went on to develop three laws of motion that govern all objects, including rockets (*see pp.126–127*). Newton also built the first reflecting telescope.

Williamina Fleming

This Scottish astronomer (1857–1911) was the leading female astronomer of her time. During her 36 years working at Harvard College Observatory in Cambridge, Massachusetts, she discovered more than 10,000 stars, catalogued several nebulae, and established the existence of white dwarf stars.

Albert Einstein

Einstein (1879–1955) is best known for his theories of relativity (*see pp.44–45*) and the equation that he made famous, $E=mc^2$, which proved that even the tiniest amount of mass can be turned into a huge amount of energy.

Edwin Hubble

This American scientist worked at the Mount Wilson Observatory in California in the early 1900s. At the time, people thought the Milky Way was the only galaxy out there. Hubble found many galaxies, and showed that the Universe is expanding (*see pp.188–189*).

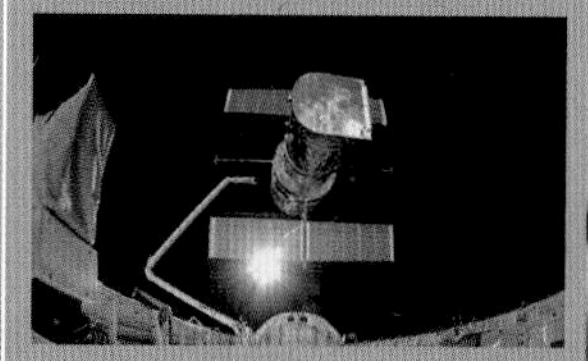

Hubble Space Telescope launching from space shuttle Discovery

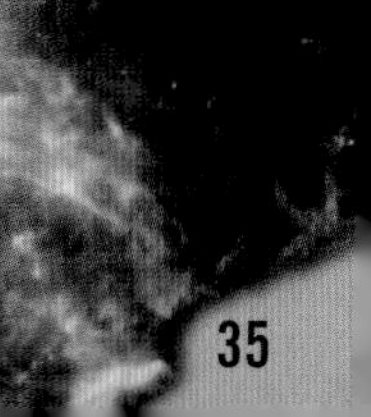

Hubble image of nebula NGC 2440

TERRAWATCH

ARTIFICIAL SATELLITES

Spies, scientists, and weather forecasters love artificial satellites because these unmanned machines see everything as they orbit Earth. You use satellites every day too, whenever you switch on the TV or radio, make a phone call, or ask the car satnav for directions. The first satellite was launched in 1957. Today, there are about 1,740 active satellites, some are as big as a school bus while others are only 4-in (10-cm) cubes.

KEY SATELLITES AND WHAT THEY DO

LOW EARTH ORBIT (LEO)

Distance above sea level: 100–1,245 miles (160–2,000 km)

Uses: spying, scientific observation, weather

Examples: Aqua, ISS, Hubble, polar-orbiting

MEDIUM EARTH ORBIT (MEO)

Distance above sea level: 1,245 miles (2,000 km)–22,236 miles (35,786 km)

Uses: Global Positioning satellites (GPS) for navigation, communications

Examples: GLONASS, Galileo

HIGH EARTH ORBIT (HEO)

Distance above sea level: above 22,236 miles (35,786 km)

Uses: weather, communications, television

Examples: Vela, IBEX

Watching the planet

Satellites do not only provide an "eye" on the planet. They monitor Earth and keep us informed about the weather, atmosphere, and climate change. Communications satellites provide a network for scientists to acquire and use this data to help prevent damage to the planet's ecosystems.

Rocky mountains

Arctic

Amazon rainforest

Atacama desert

Measuring gravity

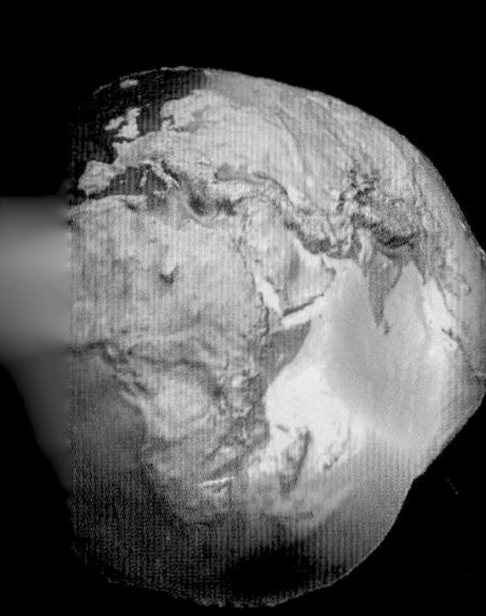

In 2005, orbiting satellites GRACE and CHAMP were used to create a map of Earth's gravitational field (*see p.45*). Known as the Potsdam Gravity Potato, red shows areas where gravity is stronger than usual, and blue where it is weaker. This helped measure changes, such as variable ocean currents and the melting of glaciers.

Below the surface

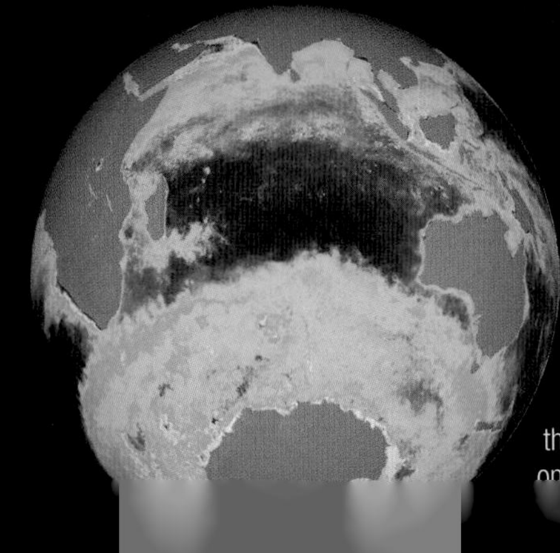

The Suomi NPP weather satellite collects data on long-term climate change and short-term weather conditions. This image shows the concentrations of chlorophyll in the southern hemisphere. Higher chlorophyll (red) shows that there is a good amount of phytoplankton, which, like plants on land, is vital as a basis for the food chain in the ocean.

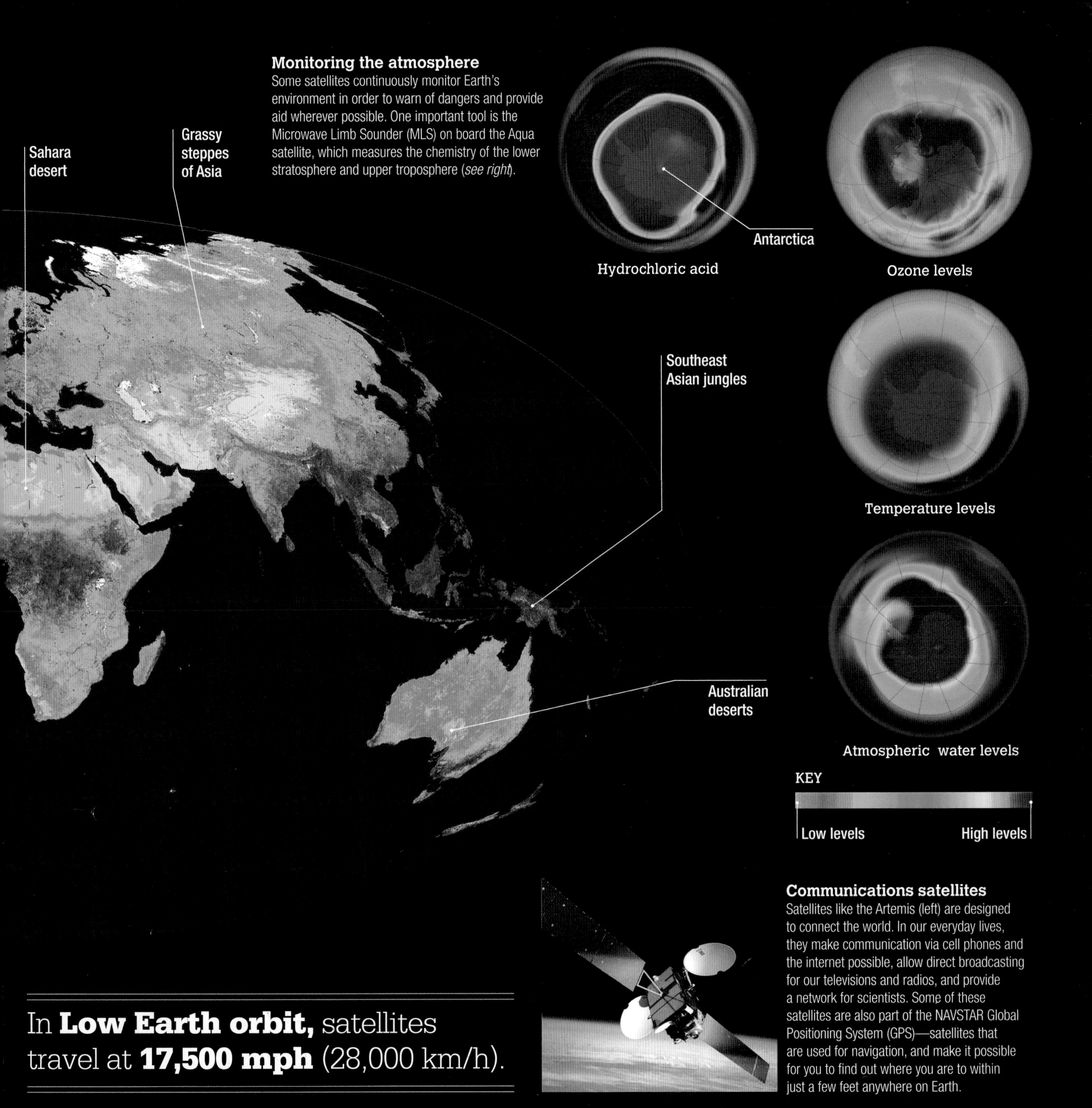

Monitoring the atmosphere

Some satellites continuously monitor Earth's environment in order to warn of dangers and provide aid wherever possible. One important tool is the Microwave Limb Sounder (MLS) on board the Aqua satellite, which measures the chemistry of the lower stratosphere and upper troposphere (*see right*).

In **Low Earth orbit,** satellites travel at **17,500 mph** (28,000 km/h).

Communications satellites

Satellites like the Artemis (left) are designed to connect the world. In our everyday lives, they make communication via cell phones and the internet possible, allow direct broadcasting for our televisions and radios, and provide a network for scientists. Some of these satellites are also part of the NAVSTAR Global Positioning System (GPS)—satellites that are used for navigation, and make it possible for you to find out where you are to within just a few feet anywhere on Earth.

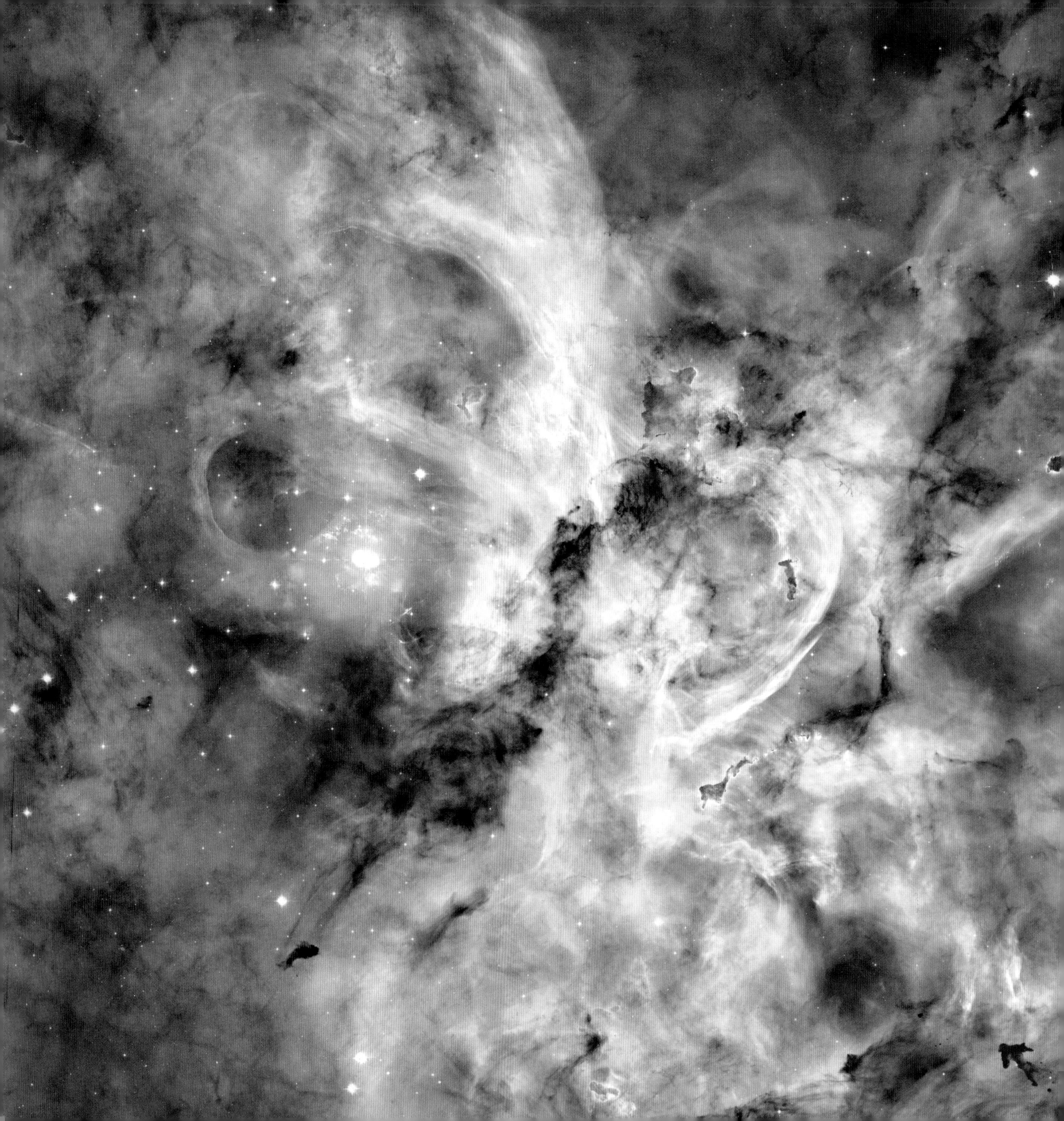

The inferno of star birth

The Hubble Space Telescope's image of the Carina Nebula shows the turbulent effects that the outflowing winds and ultraviolet radiation of newborn stars inside the nebula have on the giant cloud in which they were born. This enormous nebula is around 7,500 light-years from Earth in the southern constellation Carina (the Keel). The image has been assembled from 48 frames taken by Hubble's Advanced Camera for Surveys, and the colors added. The orange-red colors are sulfur, green is hydrogen, and blue indicates oxygen emission.

HUBBLE

SPACE TECHNOLOGY

The Hubble Space Telescope is one of the most successful scientific instruments ever built. The largest of all the space telescopes, it has made more than 1.3 million observations since its launch in April 1990, whirling around our planet at 17,000 mph (27,360 km/h). With its images of distant space, it has altered our view of the Universe and our place in it.

Original image of M100 galaxy

Image after the mirror was fixed

Astronauts spacewalking to correct the mirror

Fixing Hubble

The Hubble Space Telescope is well known for its wonderful images of space. However, it has not been without its problems. When it went into orbit in 1990, the primary mirror was out of place by less than a millimeter and had to be fixed during spacewalks.

Aperture door
This can be closed if necessary to protect the telescope from damaging light from the Sun.

Primary mirror housing

Solar panel
These produce enough power for all the science instruments to operate together.

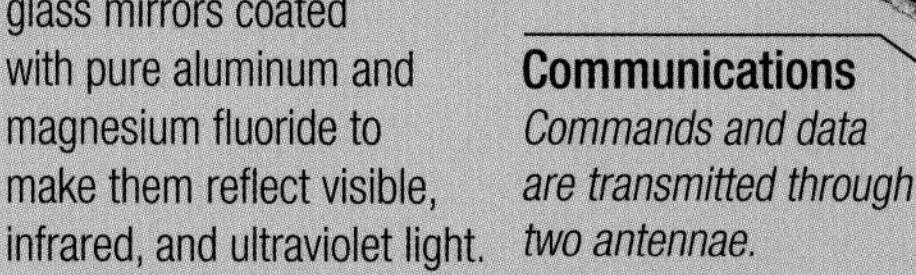

Inside Hubble

Hubble is a reflecting telescope (*see p.20*), with glass mirrors coated with pure aluminum and magnesium fluoride to make them reflect visible, infrared, and ultraviolet light.

Communications
Commands and data are transmitted through two antennae.

Pillars of Creation

In 1995, Hubble took what has now become one of the most famous pictures taken of space. The image showed towering columns in the Eagle nebula, a cloud of interstellar gas and dust 7,000 light-years from Earth that lies in the constellation Serpens. The three "pillars" are part of an active star-forming region and contain newborn stars.

Sun
Full Moon
Venus at brightest
Vega
Polaris
Naked eye limit
Hubble Space Telescope
-30 -20 -10 0 10 20 30
Brighter
Dimmer

Vega

Measuring brightness

Cameras on Hubble are regularly adjusted, or calibrated. To do this, Vega,the brightest star in the constellation Lyra, is used as the baseline (zero) for a starlight scale of brightness called the apparent magnitude scale. Objects with apparent magnitude higher than 6 are too dim for us to see.

Hubble has **looked back** into the **past**, to a galaxy that is **13.4 billion light-years** away from **Earth**.

Hubble's vision

Hubble is named after the astronomer Edwin Hubble who made the first major study of far-off galaxies. Among many discoveries, the far-seeing telescope has found two new moons orbiting the dwarf planet Pluto, and captured the 1994 collision between the comet Shoemaker-Levy 9 and Jupiter. It has also imaged the Andromeda galaxy, spotted eruptions on Jupiter's moon Europa, and found water in the atmosphere of an exoplanet, Fomalhaut b, 700 light-years away.

Casting a giant mirror

In the Chiliean Andes, the Giant Magellan Telescope is being constructed. It will combine the light from eight 8.4-meter mirrors to produce infrared images ten times sharper than the Hubble is able to produce. In November 2017, the fifth of these mirrors was cast, with 38,539 lbs (17,481 kg) of glass heated for several days, then cooled for a couple of months. This will be followed by a three-year long process of polishing.

SPACE, MIRRORS, AND TELESCOPES

BY PROFESSOR PAUL HICKSON, ASTROPHYSICIST

The twinkling of stars is caused by the bending of their light in the Earth's atmosphere, which also blurs their images. Modern telescopes can correct this. Their large mirrors collect much more light than we see with our eyes, allowing us to study very distant stars and galaxies.

The Hubble telescope's main mirror is 7.87 ft (2.4 m) across and has a 10-in (25.4-cm) thick core made of glass and hollow spaces, fused between 1.5-in (4-cm) thick solid glass plates. The gaps make it far lighter than a solid mirror, and the design allows for the expansion and contraction that happens when warm sunlight reaches the coldness of space.

But when the first images from Hubble appeared, they were disappointingly blurred. It turned out that there was a tiny flaw in the mirror of only a few nanometers, or 1/50th of the thickness of a sheet of paper. The fault was fixed in an extraordinary mission to space when a team of astronauts fitted extra mirrors to adjust the light reaching the central point.

Astrophysicist
Professor Paul Hickson has contributed to the design of telescopes on Earth and in space. In 1994, he collaborated with Ermanno Franco Borra to construct the first liquid mirror telescope.

The James Webb telescope launching in 2019 has a far larger mirror that is 21.3 ft (6.5 m) wide. Its mirrors are polished to an accuracy of about 1 millionth of an inch, and coated in a thin layer of gold to help pick up infrared light. This makes the telescope so sensitive that it needs protection even from light reflecting off Earth and the Moon, and a giant sunshield the size of a tennis court blocks the direct sunshine that would blind it.

There is no doubt that space-based telescopes such as Hubble offer a fantastic view across the Universe, but it takes a sky-high budget to launch a telescope into space. One recent advance for land-based telescopes is based on very old knowledge. When you spin a pool of liquid, its surface forms a curve where every point along the edge is an equal distance from a fixed point. This is the concept behind liquid mirror telescopes such as the 20-ft (6-m) wide mercury mirror of the Large Zenith Telescope. One day we will have space telescopes on the Moon. Maybe some of them will have at their heart not a bank of mirrors, but a spinning pool of metal.

"Orbiting **the Sun**, the **James Webb Telescope** will allow us to see **Earthlike planets** circling other **stars**."

Grinding Hubble's primary mirror

The liquid mirror of the Large Zenith Telescope, Vancouver, Canada

Assembling the Chandra space telescope

6 of the 18 mirrors for the James Webb Space Telescope

Floating without gravity

Inside a plane—a "vomit comet"—trainee astronauts float in zero gravity. In what is called a parabolic arc, the plane has climbed steeply, leveled off, and then dived at a 45 degree angle. Going over the top of the arc, both the plane and the people inside it go into freefall, accelerating at the same rate and canceling out gravity. The astronauts experience weightlessness for 30 seconds.

THE PHYSICS OF SPACE

As we travel beyond Earth through our solar system, we encounter some of the most challenging concepts of all, about the nature of ourselves and our relationship with the Universe. The most brilliant of human minds have managed to explain these, at least partly, with incredible insights. In recent times, Albert Einstein (*see p.35*) laid the foundations of modern physics with his theories about how the Universe works.

Gravity

In the late 1600s, Isaac Newton identified gravity as a force of attraction between all objects with mass. The force depends on how massive the objects are and the distance between them. Einstein proved mathematically that gravity is not so much a force but a warping, or curving, of space and time. So a massive object, such as a planet, or the Sun, generates a gravitational field by curving the surrounding spacetime.

Presence of the Sun warps space

Gravitational waves

Einstein showed that massive, accelerating objects, such as black holes orbiting each other, cause ripples in spacetime. These ripples always travel at the speed of light through the Universe, bringing with them information about their origins. Gravitational waves were not actually observed until 2015.

Plot of a gravitational wave

Gravitational waves caused by colliding black holes

Gravity and orbits

Orbits are repeating paths that one object takes around another in space, for example a planet around a star. Without gravity, the planet would go off into space in a straight line. With gravity it is pulled back toward the star.

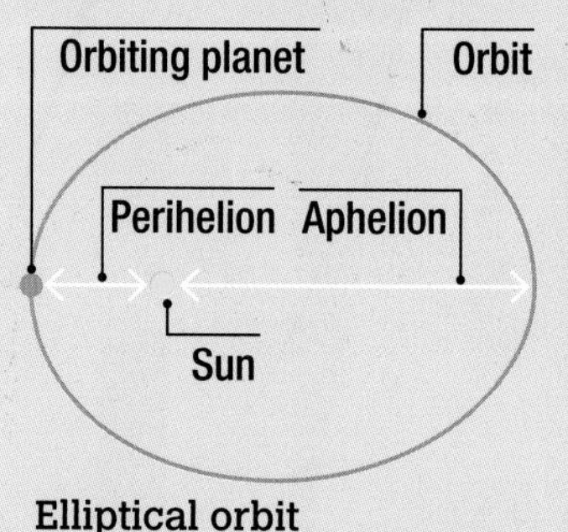

Elliptical orbit

Gravity of Earth holds the Moon in orbit

Time

Einstein realized that time does not progress at the same rate for everything, everywhere. The rate at which time flows depends on where something is and how fast it is traveling.The differences only become evident if something moves fast enough through space. The closer the speed of an object to the speed of light, the more time dilates (expands). Astronauts who stay a long time on the ISS experience time dilation.

After 803 days on the ISS, Sergei Krikalev returned 0.02 seconds younger than if he had stayed on Earth.

Light

Some stars and galaxies are so far away that the light we are seeing left them billions of years ago. That distance is measured in light-years—the distance that light travels in one year, equal to 5.9 trillion miles (9.5 trillion km).

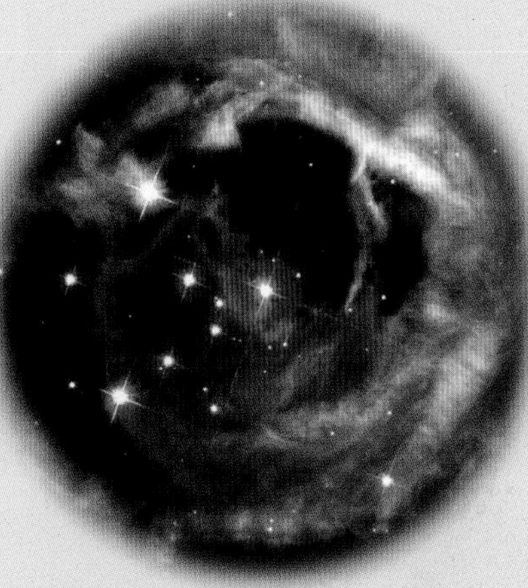

Light echo from the star V838 Moncerotis

$E=mc^2$

Nothing can travel faster than light, and the speed of light in a vacuum is always the same. Einstein showed that mass is energy and expressed his theory in the formula $E = mc^2$, where E is energy, m is mass, and c is the constant speed of light.

Light travels at **186,282 miles** per **second** (**299,792 km** per **second**).

OUR SOLAR

SYSTEM

THE SUN'S FAMILY

The solar system is just one small neighborhood among many in the vastness of space. It is home to 8 planets (4 rocky, 4 gassy), 5 icy dwarf planets, 181 moons, millions of asteroids, and billions of comets. The surfaces of many of these objects are pitted with craters blasted out by meteorites. In the middle, holding everything in place with its gravity, is a medium-sized star that we call the Sun.

PLANETARY TYPES

The four inner planets—Mercury, Venus, Earth, and Mars—are far smaller than their outer cousins—Jupiter, Saturn, Uranus, and Neptune. This is partly because the inner planets are nearer the Sun and so feel the full force of its solar winds—streams of particles blasted out by the star. These winds blew away much of the gas and dust from the closest planets as they developed, leaving smaller globes of rock and dust. The outer planets had more time to attract and retain gas and water because their gravity was competing with weakened solar winds farther from the nuclear-powered Sun. They grew into icy, gassy giants.

PLANETARY DEFINITION

Over the years, astronomers have called various other small objects orbiting the Sun planets—indeed Pluto was listed as the ninth planet for 76 years. Then, in 2006, the International Astronomical Union defined a "dwarf planet" as a round object orbiting the Sun that has not cleared the neighborhood around its orbit. Pluto lost its status as a planet and was soon joined by Ceres (originally called a planet, then an asteroid) and newly discovered Eris, the largest in a group now described as dwarf planets. Two years later Makemake and Haumea were added.

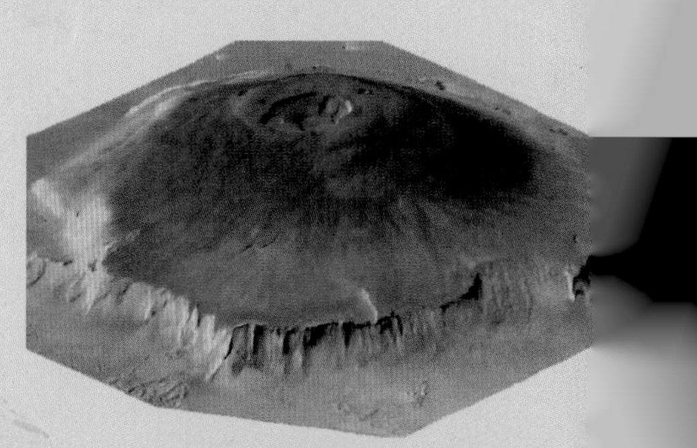

Olympus Mons volcano, planet Mars

THE SUN

YELLOW DWARF STAR

Our yellow Sun burns over half a billion tons of hydrogen a second in constant nuclear reactions in its core. This releases the energy that, when it reaches our planet, keeps us alive. Our star contains more than 99 percent of all the matter in the solar system, and its gravity—28 times stronger than on Earth—holds planets and lumps of ice and rock in place over billions of miles. This vast ball of flowing plasma is truly the ruler of the solar system.

Star profile: Yellow dwarf

Average distance from Earth: 92,956,000 miles (149,600,000 km)

Mass: (Earth = 1) 333,060

Equatorial circumference: 2,715,396 miles (4,370,006 km)

Surface temperature: 10,000°F (5,500°C)

Core temperature: 27 million °F (15 million °C)

Rotation period at equator: 25 days

Rotation period at poles: 36 days

Chemical composition: 73% hydrogen, 25% helium, 0.36% carbon, 1.64% others including nitrogen, neon, oxygen, iron, silicon, sulphur

Energy generator

In the Sun's core, hydrogen is converted to helium in nuclear fusion. In the process, energy is carried outward by light units called photons. It takes a photon up to 200,000 years to travel through the radiative and convective zones to reach the surface. Yet when the energy seeps out, it takes only 8.3 minutes to travel through space to reach Earth.

Sunspots

These areas on the surface—some larger than Earth—look dark because they are cooler. They are the areas of the strongest magnetic activity.

Corona

This is the outermost layer of the Sun's atmosphere, starting at 1,300 miles (2,100 km) above the surface.

Coronal mass ejections

These are huge explosions of plasma and magnetic field from the corona. Their effects usually take two to four days to reach Earth.

Power plant

Life on Earth would not be possible without the Sun. Its power plant energy provides heat and light so that plants can photosynthesize—make sugars for food. The plants in turn become the fuel that allows animals to live and grow.

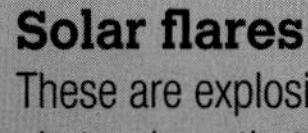

Solar flares

These are explosions in the photosphere that can heat surrounding material to millions of degrees in a few seconds. They produce bursts of x-rays, ultraviolet and electromagnetic radiation, and radio waves.

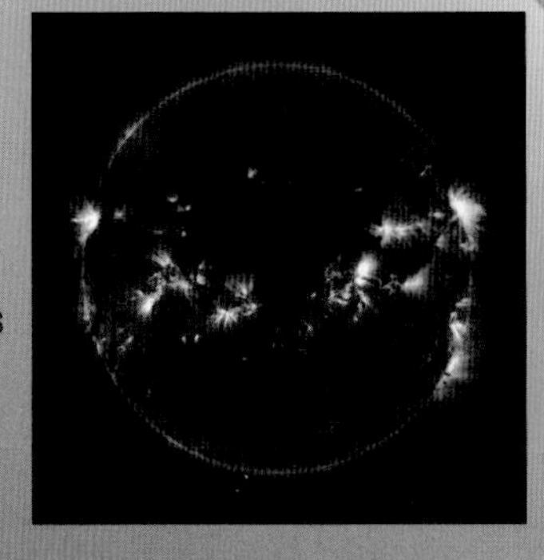

Radiative zone
The Sun's energy bounces around in very dense gas here for many thousands of years. The energy is passed over and over again between atoms in particles called photons.

Convective zone
Here, rolling currents of hot gas carry the Sun's energy in the form of photons outward to the surface.

Photosphere
This is the visible surface of the Sun. It is 250 miles (400 km) deep and is where the Sun's energy is released as light.

Chromosphere
This layer of the Sun's atmosphere is 250–1,300 miles (400–2,100 km) above the photosphere. It has a rosy red color that can be seen during eclipses.

Coronal holes
These dark areas in the corona are cooler and scientists think they are areas where particles of the solar wind escape.

Light travels at **186,000 miles per second (299,792 km/s)**.

Sun to Earth
1AU (8.3 light-minutes)

Sun to Mars
1.5 AU (12.6 light-minutes)

Sun to Proxima Centauri
4.3 light-years

Sun to far side of Milky Way galaxy
77,000 light-years

Time and distance

The distance between the Sun and its planets is measured in Astronomical Units (AU). The distance between the Sun and the Earth is 1AU and other distances are measured against this one *(see pp.54–55)*. A light-year is the way astronomers describe how far light can travel in one year—a distance of 6 trillion miles (10 trillion km).

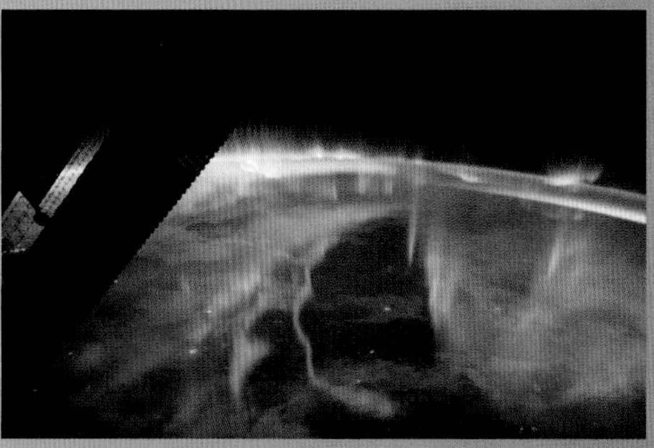

Earth's dancing lights

Giant clouds of gas from a coronal mass ejection collide with Earth's magnetic field, dislodging particles which interact with gases in the atmosphere and cause the dazzling curtains of light that are an aurora. The aurora borealis is at the North Pole, and the aurora australis at the South Pole.

Seeing the invisible

Techniques have been developed to image the Sun that reveal information that cannot normally be seen for scientists to analyze.

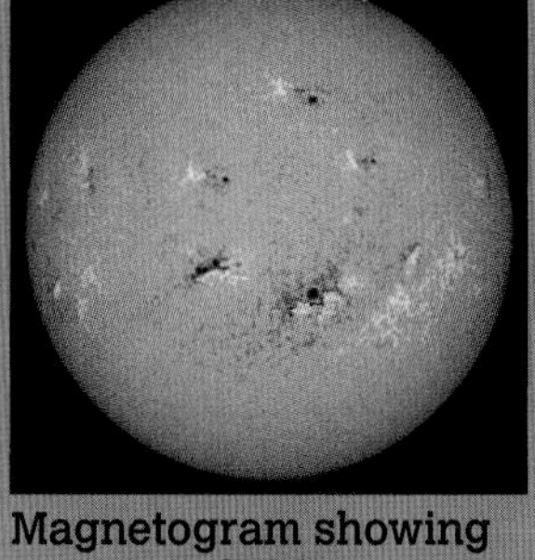

Magnetogram showing magnetic fields

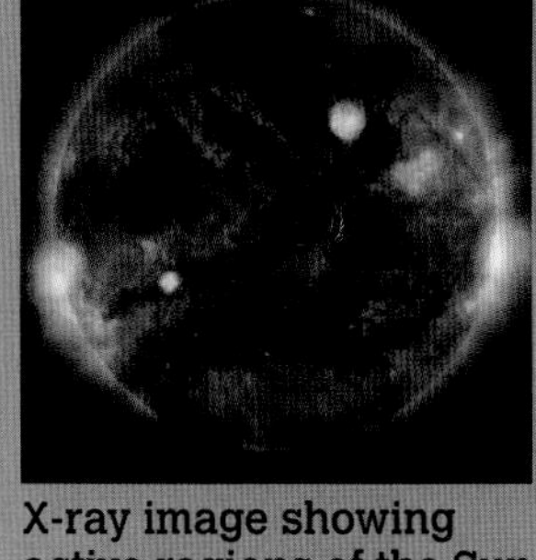

X-ray image showing active regions of the Sun

Extreme ultraviolet image showing ionized gas surrounding the Sun

Sun god

The importance of the Sun to everyday life has been recognized by different cultures. Here, the ancient Egyptian sun god Ra is shown traveling through the sky on his solar boat Atet, providing light to the world.

Solar mission

In 2018, a spacecraft will travel into the Sun's corona, some 3.9 million miles (6.27 million km) above its surface. NASA's Parker Solar Probe will study magnetic fields, plasma and energetic particles, and the solar wind.

Birth of our solar system

Around 4.6 billion years ago, a nebula began to collapse under its own gravity and spin faster, a dense mass forming at its center. When the mass reached 18 million °F (10 million °C), nuclear fusion began and our star, the Sun, was born. From the material held by the Sun's gravity, solar winds blew away lighter elements leaving heavy rocky materials to form planets, asteroids, comets, and moons. Farther away, the lighter elements coalesced into gas giants.

THE SUN'S PLANETS

PLANETARY DATA

We inhabit one of eight rocky or gassy planets that endlessly journey around the Sun in oval-shaped circuits. The first planet, Mercury, takes the shortest time—88 days—to complete its orbit, while the farthest, Neptune, takes the longest at 165 years. Some of these planets have satellite moons, and countless lumps of rock and ice also hurtle around our solar system.

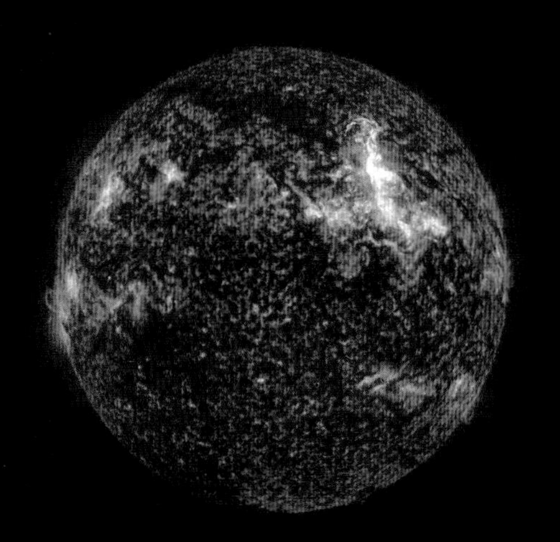

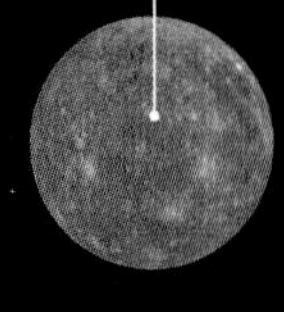

Mercury
Average distance from the Sun: 36 million miles (58 million km)
Orbital period: 88 days

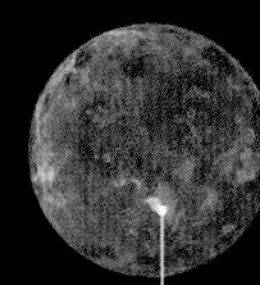

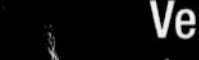

Venus
Average distance from the Sun: 67 million miles (108 million km)
Orbital period: 225 days

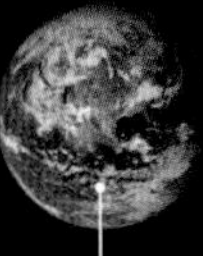

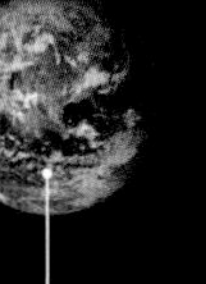

Earth
Average distance from the Sun: 93 million miles (150 million km)
Orbital period: 365 days

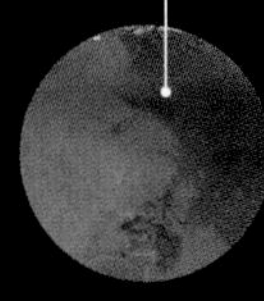

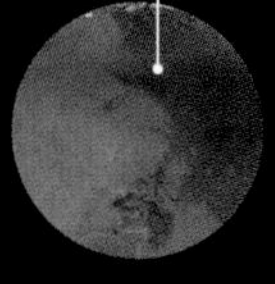

Mars
Average distance from the Sun: 142 million miles (228 million km)
Orbital period: 687 days

Jupiter
Average distance from the Sun: 484 million miles (778 million km)
Orbital period: 4,333 days

Planetary orbits

The solar system's eight planets all orbit the Sun in a counter-clockwise direction if viewed from above the Sun's north pole. They take a curved route that is affected by the Sun's gravity and that eventually brings them back to their starting points. Each revolution is a single orbit.

How far out

The chart below shows the distances of the eight planets and the Asteroid Belt from the Sun. The distances are to scale and are in Astronomical Units (AUs). 1 AU is equal to about 93 million miles (150 million km).

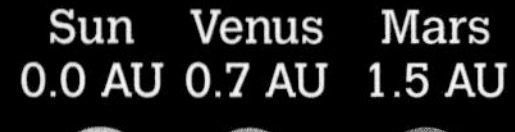

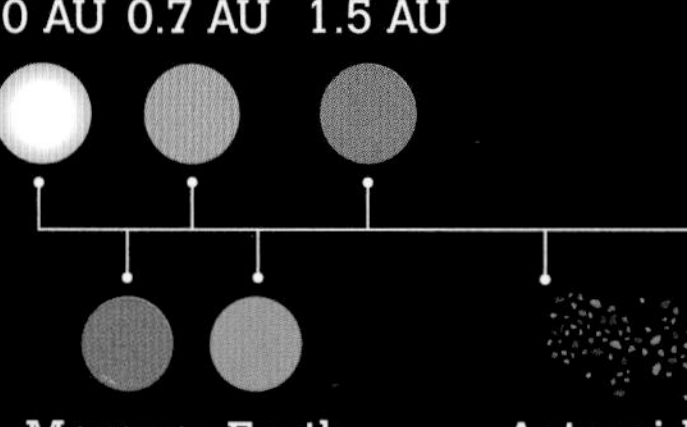

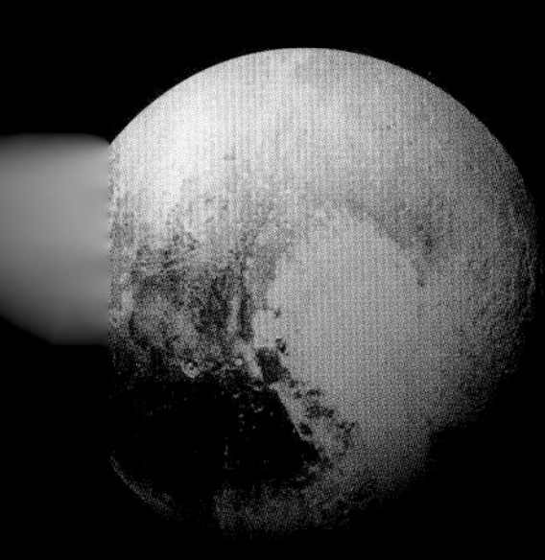

A planet no more

For 76 years from its discovery in 1930, Pluto (*see pp.114–115*) was described as the ninth planet from the Sun. However, in 2006, the International Astronomical Union, a group of professional astronomers, wrote a new definition of the term "planet" that excluded Pluto, which was reclassified as a dwarf planet.

New kid on the block

Astronomers now think that there is a planet in the outer reaches of the solar system appoximately 20 times farther from the Sun than Neptune. It is thought to be ten times more massive than Earth and that its influence may, over time, have tilted the planets of our solar system by as much as 6 degrees.

Relative sizes

The Sun, our star, is impressively big! Its radius is 10 times that of the largest planet, Jupiter, and 100 times that of Earth. More than 1,300 Earths would fit inside Jupiter and an amazing one million Earths would fit inside the Sun. Comparative diameters of the Sun and planets are shown below.

Sun 864,000 miles (1,390,000 km)

Mercury 3,032 miles (4,879 km)

Venus 7,521 miles (12,104 km)

Earth 7,926 miles (12,756 km)

Mars 4,221 miles (6,792 km)

Jupiter 88,846 miles (142,984 km)

Saturn 74,898 miles (120,536 km)

Uranus 31,763 miles (51,118 km)

Neptune 30,775 miles (49,528 km)

Uranus
Average distance from the Sun: 1.78 billion miles (2.87 billion km)
Orbital period: 30,687 days

Saturn
Average distance from the Sun: 886 million miles (1.43 billion km)
Orbital period: 10,756 days

Neptune
Average distance from the Sun: 2.8 billion miles (4.5 billion km)
Orbital period: 60,190 days

Uranus
19.2 AU

Neptune
30 AU

MERCURY

ROCKY PLANET

Blasted by solar winds that sweep away its atmosphere, Mercury's wrinkled and cratered face shows how little protection it has from the Sun or any lumps of rock and ice that smash into it from space. The smallest of our planets, it is also the second most dense—after Earth—because it is made of rock and heavy metals.

Copernicus prediction

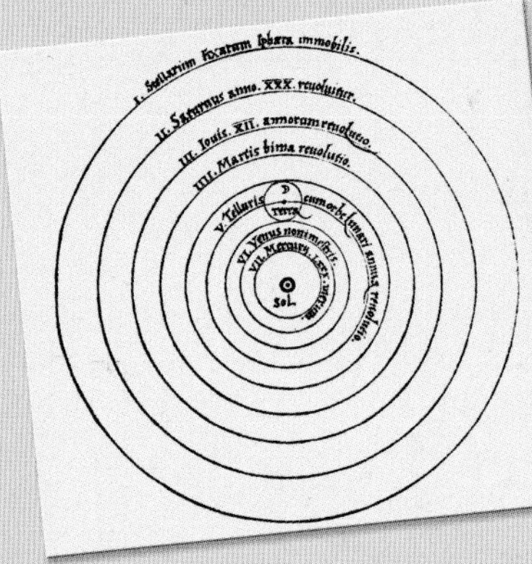

In 1543, Nicolaus Copernicus published his model of a Sun-centered solar system, concluding that both Earth and Mercury were planets. The theory was confirmed in 1610 by Galileo, and in 1639, Giovanni Zupus observed the phases of Mercury, proving that it orbits the Sun.

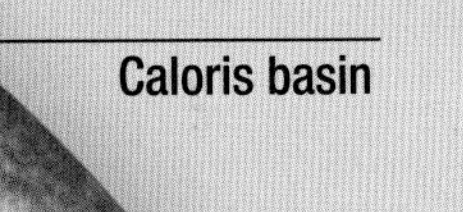

Caloris basin

Average distance from the Sun: 36 million miles (58 million km) / 0.39 AU

First record: 14th century BCE, Assyrian astronomers

Visited by: Mariner 10 (1974–1975), Messenger (2010–2015)

Speed of orbit round the Sun: 106,000 mph (171,600 km/h)

Equatorial circumference: 9,525 miles (15,329 km)

Gravity: 38% of that on Earth

Year: 88 Earth days

Peak daytime temperature: 801°F (427°C)

Lowest nighttime temperature: –279°F (–173°C)

Exosphere: very thin, mostly oxygen, sodium, hydrogen, helium, potassium

Known moons: none

Mercury's transit

Around 13 times a century Mercury passes between Earth and the Sun in the "transit of Mercury." The planet becomes visible as it travels, a tiny black dot against the golden disk of the Sun. Mercury and Venus are the only planets that lie between Earth and the Sun, so they are the only transits seen from Earth.

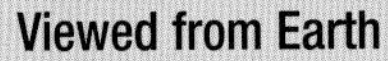

Viewed from Earth
Mercury's last transit took place on May 9, 2016. Its tiny dot (below left on the disk) is the view as it was seen from Pennsylvania.

It's all in a name

The planet is named for the Roman messenger of the gods, whose winged sandals were said to give him great speed. This seems fitting as Mercury zooms around the Sun every 88 days—much faster than any other planet in our solar system—traveling at nearly 106,000 mph (169,600 km/h).

Planetary composition

The molten iron core of Mercury takes up an extraordinary 85 percent of its volume. Its solid silicate crust and mantle cover a solid iron sulfide outer core layer, a deeper liquid core layer, and a possible solid inner core. The outer shell is only about 250 miles (400 km) thick. This false color mosaic showing different rock types was made up of images sent back by the Messenger probe. The large circular area is the Caloris basin, with its volcanic plains.

Discovery of water ice

This radar image of Mercury's north pole is shown superimposed on a mosaic of Messenger images of the same area. Despite Mercury's hot temperatures, it appears that there is ice scattered across the surface in the planet's northern regions.

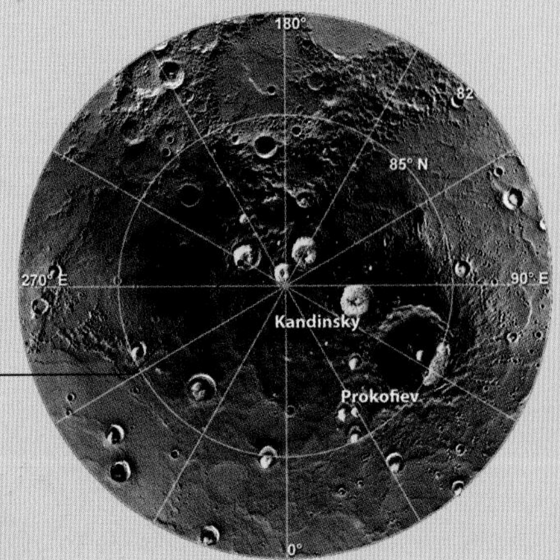

Icy hollows
It is thought that the areas indicated in green are deposits of water ice inside craters or shadowed areas.

One day on planet **Mercury** is the equivalent of **176 days** on **Earth**.

The ice beneath

This thermal map of Mercury's north pole shows dark blue areas that indicate ice. As the surface can reach temperatures of 500°F (260°C), scientists believe the ice is in permanently shaded craters under a layer of a black organic substance that has yet to be identified.

Features of Mercury

Distinctive landforms on the planet include hollows formed by impacts and lobate scarps that show the crust may have contracted.

Volcanic vent northeast of the Rachmaninoff basin

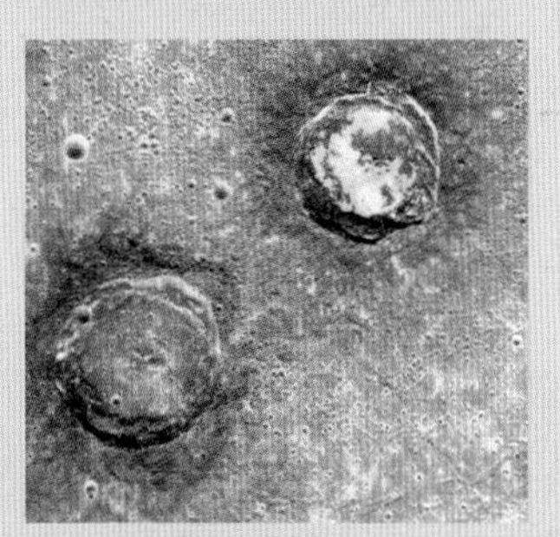

Distinctive hollows in the Caloris basin

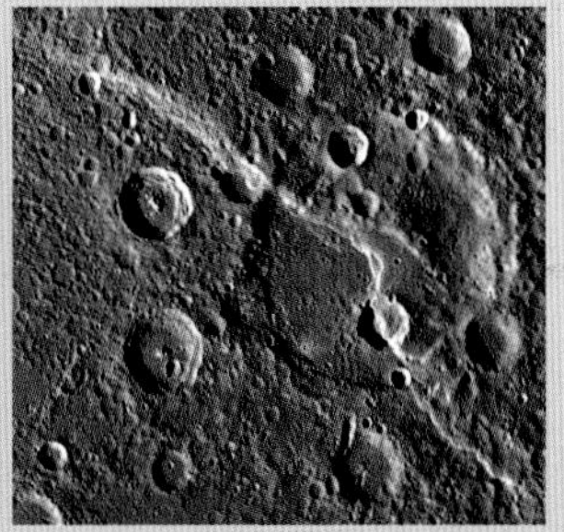

Giant scarp cutting through the Duccio crater

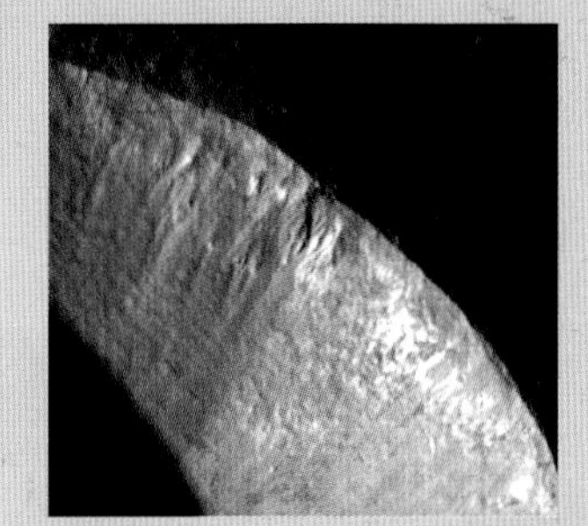

The sheer wall of a crater caused by meteorite impact

Li Po crater
This crater was named for the Chinese poet Li Bai.

Lermontov crater
This has a large circular rim.

Hokusai crater
This rayed impact crater was discovered in 1991.

Rachmaninoff basin
A double-ring basin that was imaged in a Messenger flyby.

Kuiper quadrangle
This is a heavily cratered area.

Debussy crater
The rays from this crater extend for hundreds of miles.

Pummeled from space

Mercury's solid surface is covered with craters. The planet's thin atmosphere does not break up meteorites before they hit, and without liquid water on the surface, the craters do not erode over time.

Mercury flyby reveals volcano

As the probe Messenger approached Mercury in March 2009 for its third and final flyby prior to going into orbit round the planet in 2011, its cameras took a series of high-definition images that were put together to make one picture. This shows the bright region of the Rachmaninoff basin and the nearby depression of a possible explosive volcanic vent. Smooth plains, thought to be the result of earlier volcanic activity, cover most of the area around it.

Getting ready for launch

In February 1989, solar panels were attached to the Magellan spacecraft in preparation for its launch aboard the space shuttle Atlantis on May 4, 1989. On this mission, Magellan achieved two firsts—it was the first interplanetary spacecraft deployed from a space shuttle and also the first sent to complete a radar map of Venus.

PLANETARY PROBES

Astronauts are high maintenance—they need rest and food and want to return to Earth. Space probes are far less fussy! These robot spacecraft carry out experiments and observations while they hurtle through our solar system on long, one-way missions in evermore extreme environments. They use gravitational forces to slow down or speed up, and radio back insights into worlds that we will never be able to visit.

Mariner 9

In a race to beat Soviet Mars 2, this US probe arrived first, orbiting Mars in 1971. However, a massive dust storm hid the entire surface, so Mariner 9 began by taking the first close-up images of Phobos and Deimos, the planet's two moons instead. Later it mapped 85 percent of the surface.

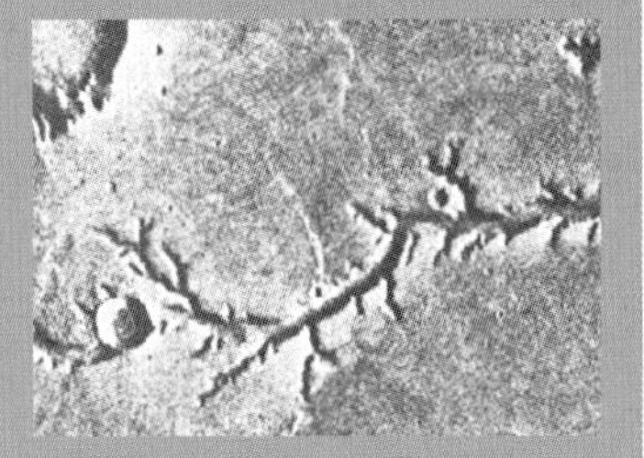

A view of channels on Mars that came from NASA's Mariner 9 orbiter

Pioneer 10

Launched in 1972, this probe was the first to voyage beyond Mars and through the Asteroid Belt. It traveled past Jupiter in 1973, and headed toward the outer reaches of the solar system. Its last, weak signal was received in January 2003.

Pioneer 10 on a Star-37E kick motor just prior to being made ready for launch

Venera 9

This unmanned Soviet space mission to Venus was launched on June 8, 1975. Its lander was the first to photograph the planet's surface while the orbiter surveyed cloud circulation.

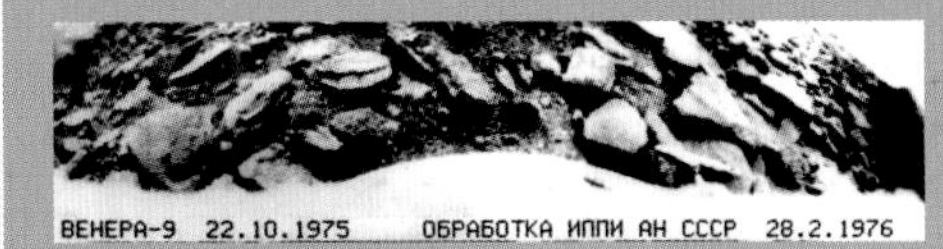

An image of the rocky Venusian surface taken by Venera 9's lander

Voyagers 1 and 2

Both probes were launched by NASA in 1977. Voyager 1 has traveled farther than any other probe—it has now moved into interstellar space. Voyager 2 is the only probe to have visited all four of the giant gas planets. Both probes carried Golden Records containing images, sounds, and greetings in 55 languages.

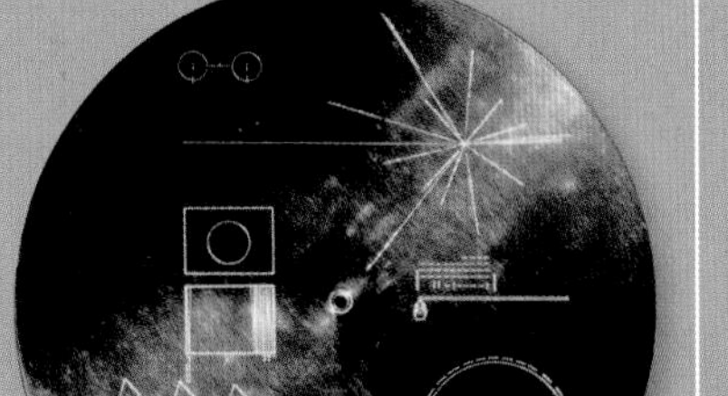

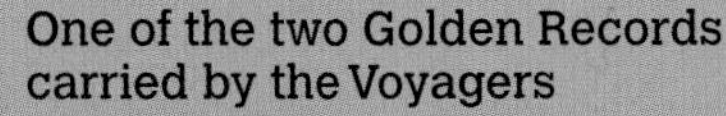

One of the two Golden Records carried by the Voyagers

Magellan

This probe arrived at Venus on August 10, 1990, and mapped the planet for three 243-day cycles—243 days being the time it took for the planet to rotate under Magellan's orbit. It was the first interplanetary mission to be launched from a space shuttle.

Magellan being deployed by the space shuttle Atlantis in 1989

Vikings 1 and 2

Launched in 1975, both of these identical probes are part of the NASA Mars Exploration mission, and both went into orbit around Mars. The Viking 1 lander touched down on the Chryse Planitia (Plains of Gold), while the Viking 2 lander is on the Utopia Planitia. Both landers took photos, collected data, and carried out experiments looking for signs of life, but found no living microorganisms.

A self-portrait of Viking 2 on Mars' Utopia Planitia

Galileo

On this NASA mission to Jupiter, the Galileo probe successfully studied the planet and its moons from 1995 until 2003. On its way there, it became the first spacecraft to visit an asteroid.

Closeup view of the surface of Jupiter's moon Europa

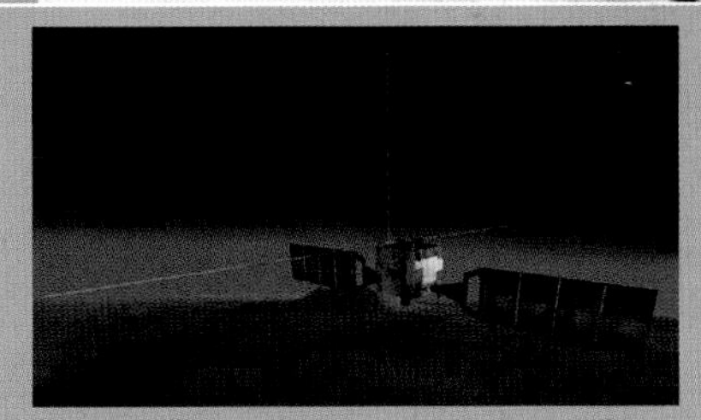

Artist's impression of Mars Express over the planet's surface

Mars Express

This was ESA's first mission to the red planet in 2004. The lander, Beagle-2, was lost as it tried to land, but the orbiter has sent back great data about Mars and its two moons.

VENUS

ROCKY PLANET

An astronaut on Venus would notice the Sun travels backward across the sky, as the planet spins clockwise. They would not have long to marvel—the acidic air, blazing heat, and crushing atmospheric pressure would leave them wheezing, roasted, and flattened before they were coated in molten lava.

Average distance from Sun: 67 million miles (108 million km) /0.72 AU

First record: known to ancient civilizations

Main visits by: Mariner 2 (1962), Venera 7 (1970), Mariner 10 (1974), Pioneer Venus Orbiter (1978–1992), Venera 13 (1982), Magellan (1990–1994), Venus Express (2006–2014), Akatsuki (2016–2017)

Speed of orbit round the Sun: 78,339 mph (126,074 km/h)

Equatorial circumference: 23,627.4 miles (38,024.6 km)

Gravity: 90% of that on Earth

Year: 225 Earth days

Surface temperature: 864°F (462°C)

Atmosphere: carbon dioxide, nitrogen, sulfuric acid

Known moons: none

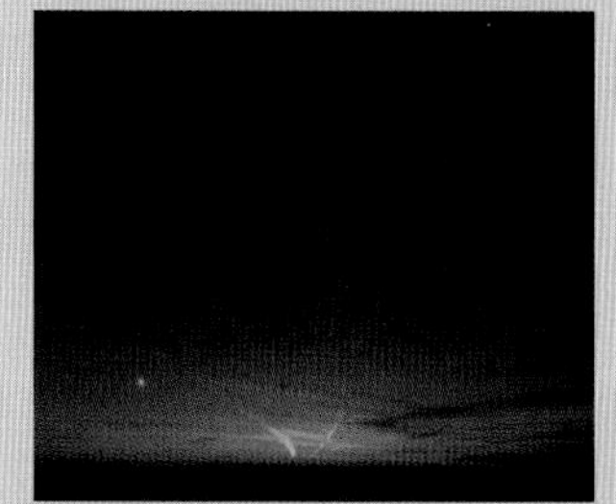

Morning "star"
Venus is sometimes called the morning star, although it is not a star at all. It can be seen clearly just before sunrise and at this time of day is 19 times brighter than the brightest star in the night sky, Sirius.

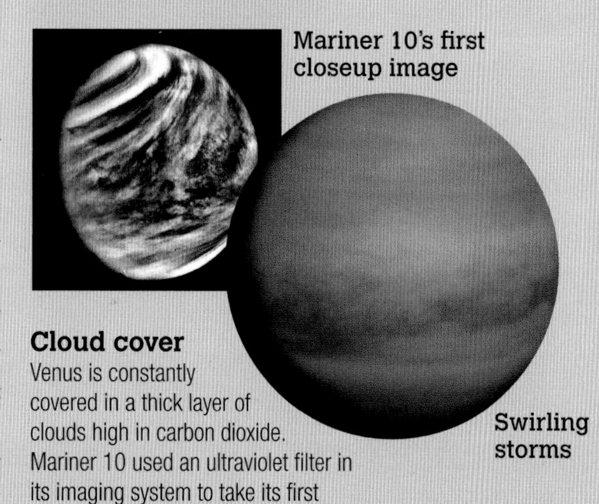

Mariner 10's first closeup image

Cloud cover
Venus is constantly covered in a thick layer of clouds high in carbon dioxide. Mariner 10 used an ultraviolet filter in its imaging system to take its first pictures of the planet.

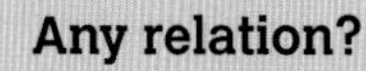

Swirling storms

Beauty
Venus is the only planet named for a female—the Roman goddess of love and beauty. The planet shone the most brightly of the five planets that were known to ancient astronomers.

Any relation?
Venus is sometimes called Earth's sister because it has a similar size and mass. But its air is mostly heat-trapping carbon dioxide (the greenhouse effect gone mad) and there are no seasons because Venus has virtually no tilt. Its surface has huge smooth plains formed by the lava that flowed from ancient volcanic eruptions. We know this because, in the 1990s, the Magellan spacecraft used an imaging radar technique to show what the surface was like under the clouds (left). Scientists were then able to produce a false color map (right), in which red represents mountains and blue represents plains.

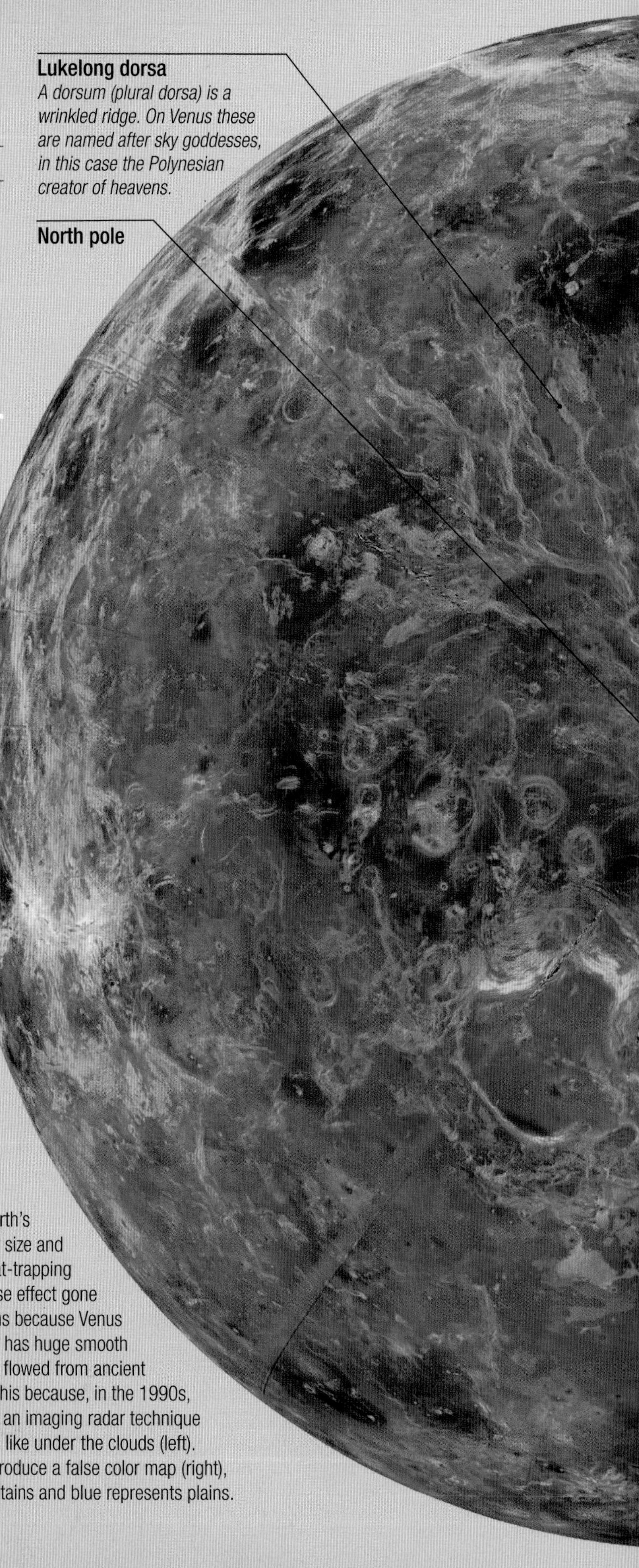

Lukelong dorsa
A dorsum (plural dorsa) is a wrinkled ridge. On Venus these are named after sky goddesses, in this case the Polynesian creator of heavens.

North pole

Atalanta planitia
About 80 percent of the surface is covered by smooth, volcanic plains. This large, broad depression is a lowland area that shows signs of having been flooded with lava in ancient volcanic flows.

Ishtar Terra highlands
One of two "continents" on Venus, Ishtar Terra is similar in size to Australia. Aphrodite Terra, on the equator, is twice the size.

Maxwell Montes massif with its peak, Skadi Mons

A **year** on **Venus** is **shorter** than a **day** on the **planet**.

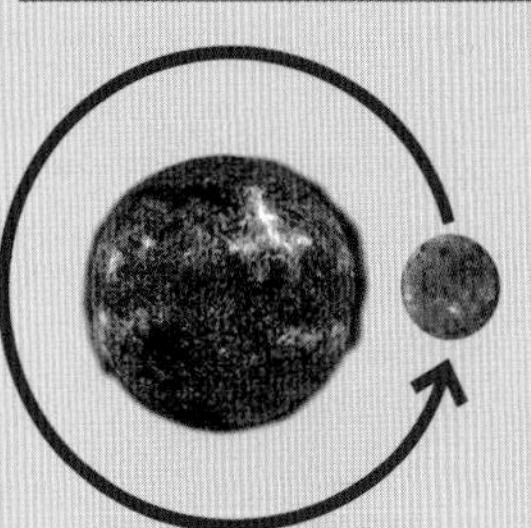

243 Earth-days =
1 rotation of Venus =
1 day on Venus

225 Earth-days =
1 orbit of the Sun =
1 year on Venus

Slow rotation

Venus spins very slowly on its axis and this accounts for the fact that a single rotation period, or "sidereal day," takes 243 Earth days. This is longer than it takes the planet to orbit the Sun, a period of 225 Earth days.

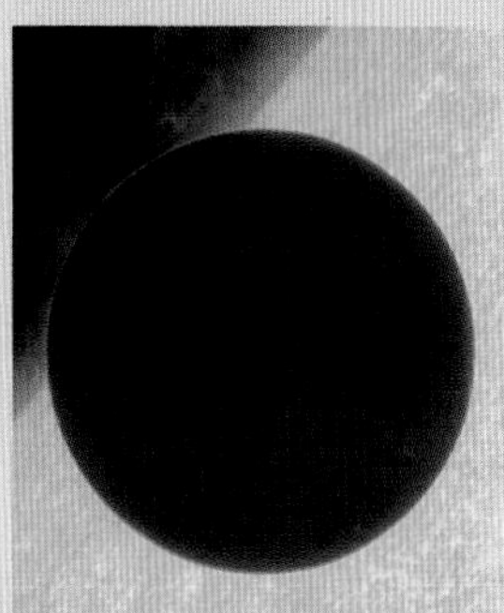

Transit

This photo was taken in 2012 by the Japanese Hinode spacecraft. It captured the transit of Venus as the planet passed across the face of the Sun. It occurred as one of a pair of rare transits. The next transit will happen in 2117.

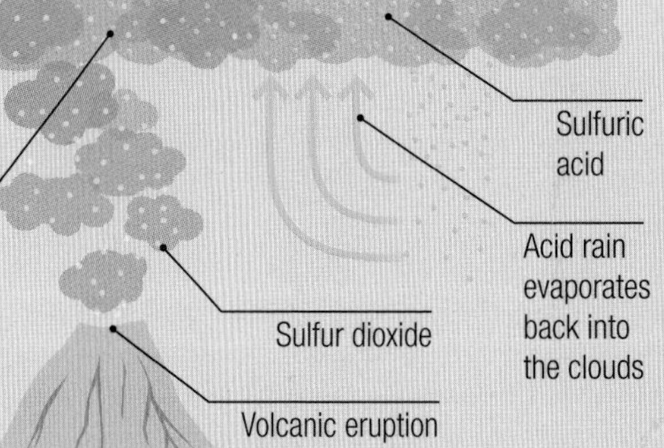

Acid rain

It rains sulfuric acid on Venus. Thick layers of cloud 30 miles (50 km) deep swathe the planet. They consist mainly of sulfur dioxide and sulfuric acid and are caused by volcanic eruptions on the surface. This acidic rain evaporates in intense heat before it can reach the surface.

Features of Venus

Despite its sulfurous clouds, probes have sent back many images of the surface of Venus.

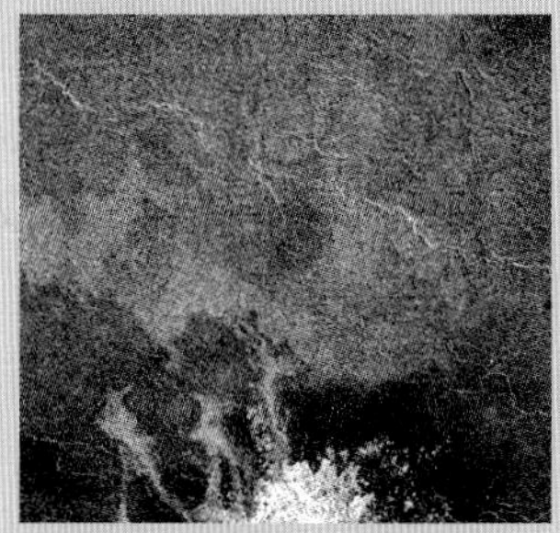

Lava flows from shield volcano Sif Mons

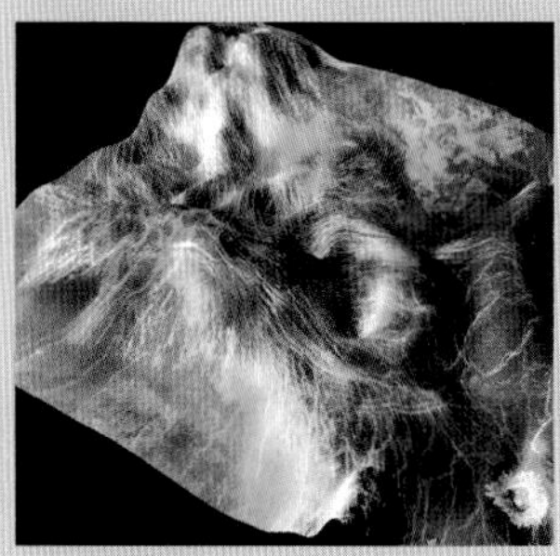

Magellan images of plains in Sedna Planitia

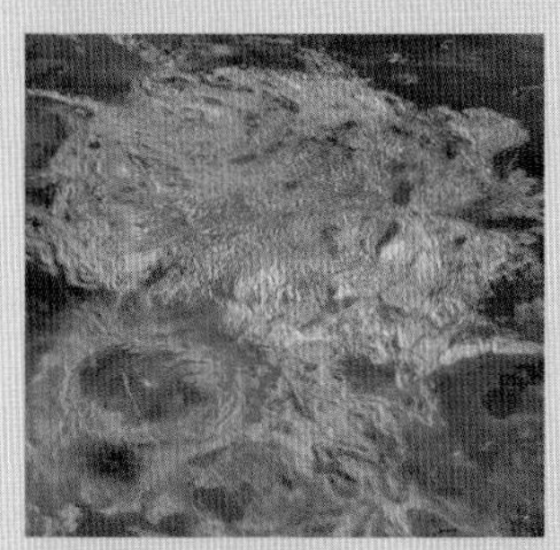

Southern plateau highland Alpha Regio

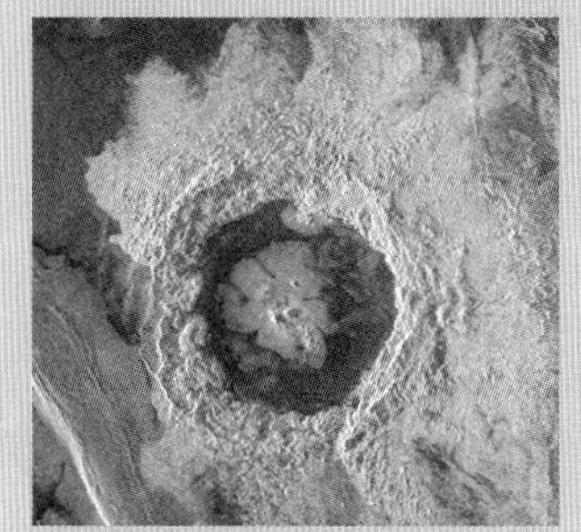

Impact crater Dickinson, 43 miles (69 km) wide

Volcanic activity on Venus

In this computer-generated image of Venus, based on data from the Magellan probe, the volcano Sapas Mons rises up from the hot, desolate landscape of the planet. Lava flows extend for hundreds of miles over the plain. This is just one of over 1,600 major volcanoes to be found on Venus, more than any other planet in our solar system. In 2014, data from the Venus Express orbiter convinced scientists that some of the volcanoes may be much younger than the rest of the surface, which has an estimated age of 300–500 million years.

Blue planet

Earth is a fragile planet with a surface of thin plates floating on a hot rocky mantle above a boiling, liquid outer core. The planet sits near enough to the Sun to bask in its warmth without burning. There are seasons because of its tilt (caused by a massive collision), and weather because of an atmosphere that holds water vapor.

EARTH

ROCKY PLANET

Stand on the equator and you are spinning with the planet at around 1,000 mph (1,610 km/h), but at the North or South pole, you are quite still—except you are zooming along at 67,000 mph (108,000 km/h), the speed of Earth orbiting the Sun. Meanwhile, Earth's magnetic field helps to deflect solar winds, and the ozone layer blocks solar radiation.

Average distance from Sun: 93 million miles/150 million km/1 AU

Speed of orbit round the Sun: 66,622 mph (107,218 km/h)

Equatorial circumference: 24,872.6 miles (40,030.2 km)

Peak daytime temperature: 136°F (58°C)

Lowest nighttime temperature: –126°F (–88°C)

Atmosphere: includes nitrogen, oxygen, argon, carbon dioxide

Known moons: 1

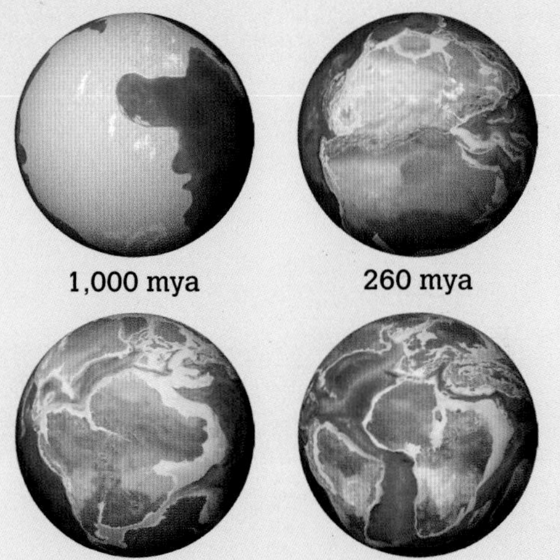

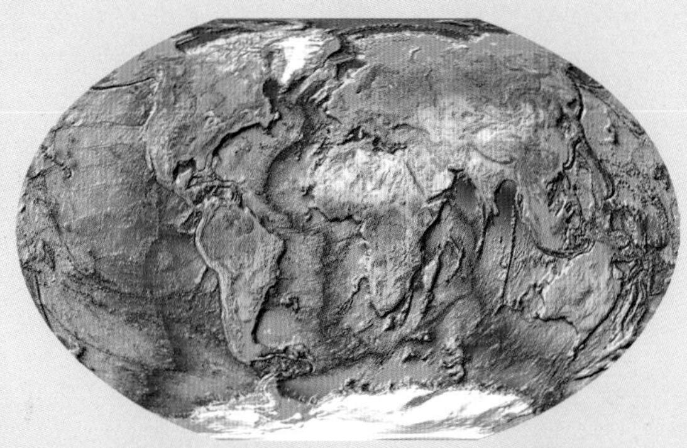

Present-day surface topography

Shifting scene

Earth's surface changes, moving on the molten layer beneath, and over time causing mountains to rise and continents to fracture. Today, there are seven large tectonic plates as well as many smaller ones. Their movements cause earthquakes and volcanic activity that influence Earth's atmosphere and oceans.

Earth's magnetic field

Aurora borealis at North Pole

Magnetic field

Movements of molten metal deep inside Earth generate a magnetic field, or magnetosphere. This protects the planet from harmful radiation from space, sometimes creating aurorae at the poles.

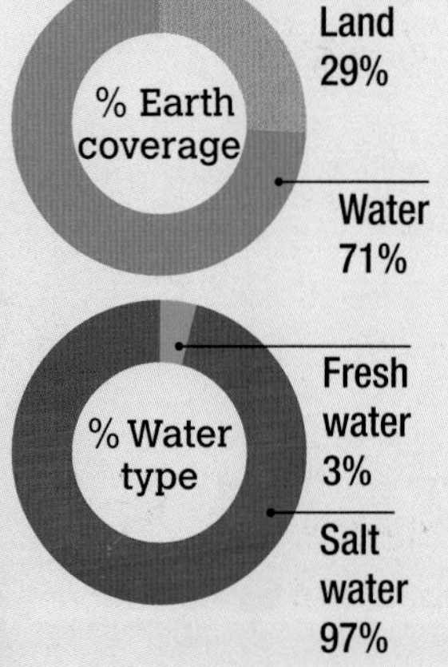

Life started in the seas, and plant life produced oxygen that fostered more.

Designed for life

Our "Blue Planet" is so-called after its nitrogen-rich atmosphere and abundant oceans that do not freeze or evaporate because the temperature keeps them mostly liquid. The water rests on a thin crust that floats on a mantle of flowing hot rock. All this is key to why Earth is the only planet known to have life.

Features of Earth

Views of the planet's surface reveal some extraordinary sights, from volcanoes to hurricanes.

Mauna Loa, the planet's biggest volcano

San Andreas Fault tectonic boundary

Chapman Glacier in Antarctica

Hurricane storm zone in the Atlantic Ocean

SPACE DEBRIS

FAST-MOVING OBJECTS

Space debris is a growing problem. More than 5,000 launches have sent up many tons of equipment that can break down, get damaged, or be blown apart by leaking fuel. Some of this junk stays floating around the planet at such a speed that even a fleck of paint can crack the window of a spacecraft.

Types of debris

Space debris includes both natural and artificial objects. About 14 percent of tracked objects are satellites, of which less than a third are operational. About 18 percent are upper stages of rockets and other objects from missions, from lens covers to a toothbrush.

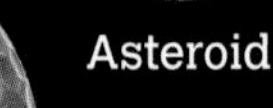

Asteroid

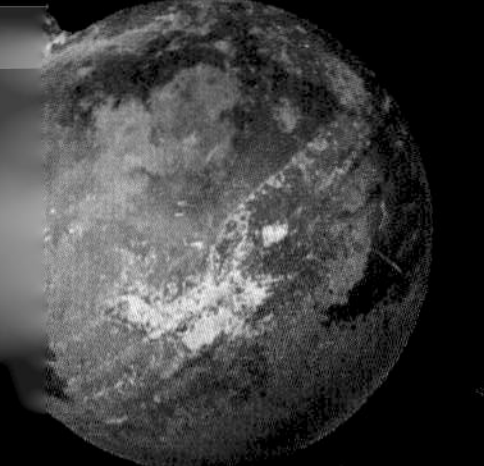

Parts of satellites and space vehicles

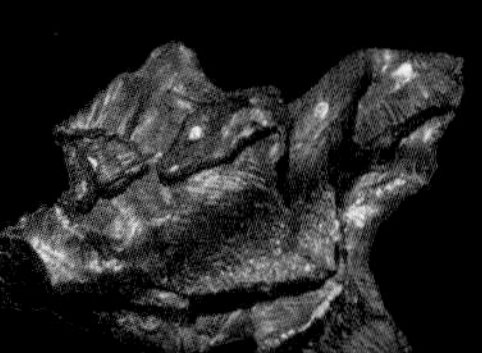

Bits of deformed metal

Made-for-space 0.1 oz (3 g) golf ball

Danger in space

Damage has already been done to space stations. Paint flecks only a few mm wide caused a large chip in a window in the ISS in 2016. Its massive 31-in (80-cm) wide windows resisted the impact, but something larger than 4 in (10 cm) could have caused critical damage.

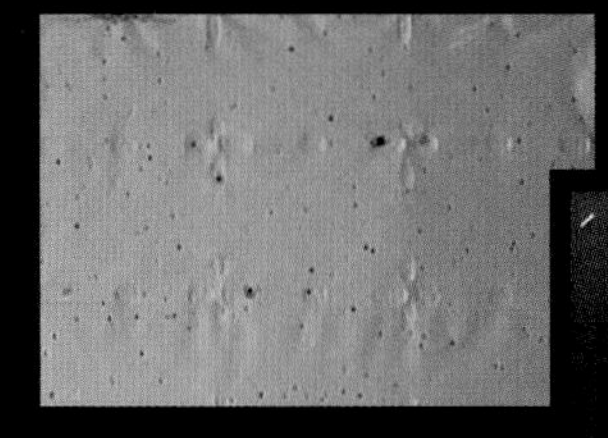

Small holes in a satellite panel

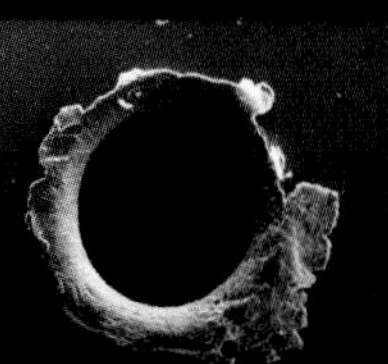

Micrograph detail of small object damage

Protecting spacecraft

Space organizations are investigating ways to protect working spacecraft and clean up space. The European Space Agency's Clean Space initiative plans to launch "e.Deorbit" in 2023 to track down a derelict satellite 480 miles (770 km) above Earth. The vehicle will steer the satellite back into the atmosphere to burn or break up there under control.

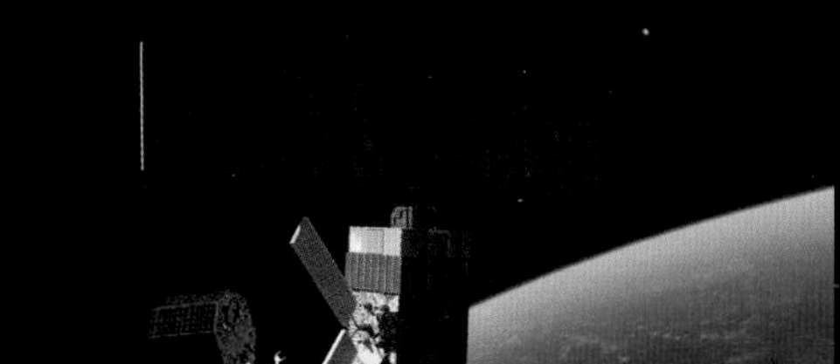

Payloads
Red dots represent parts of loads carried by spacecraft.

Personal loss
Debris includes tools and a pair of space gloves.

Watch out!
In 1997, a woman in Oklahoma was hit on the head by part of a booster from a shuttle, but she was not hurt.

Earth surrounded

This is a computer image showing the horrifying number of known pieces of human-created space debris over 60 in (150 cm) in size orbiting Earth. The amount has become a real and dangerous issue, so scientists are looking at ways to clean up the accumulated mess of 60 years of space travel. The density of the objects means that there is an increasing danger that working satellites will collide with the orbiting junk and generate yet more debris.

In 2013 there were more than **20,000 pieces** of debris **larger than a softball** orbiting the Earth.

Debris reaches the screen

Scientists are not the only ones thinking about the threat posed by space debris. The 2013 film *Gravity* begins with the devastating effect debris from a destroyed satellite has on a space station and its astronauts. This is science fiction, but in January 2007, the Chinese destroyed one of their satellites, increasing trackable debris by 25 percent.

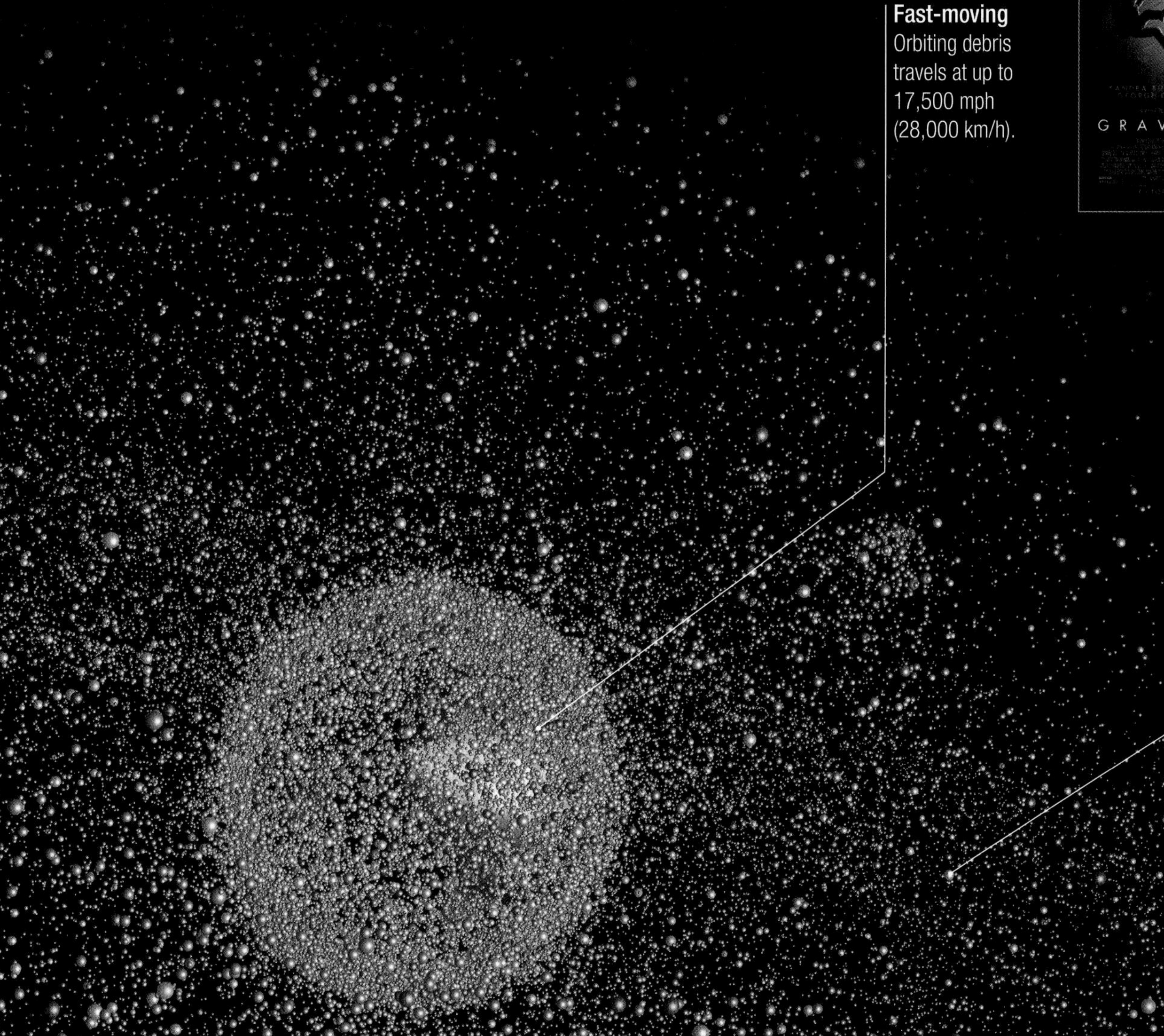

Fast-moving
Orbiting debris travels at up to 17,500 mph (28,000 km/h).

Rocket parts
Yellow dots are the upper stages of rocket bodies.

Fragments
Blue and white dots represent tiny pieces of metal and other debris.

EARTH'S MOON

NATURAL SATELLITE

Moons are space bodies that orbit a planet. Earth has one spinning 238,857 miles (384,403 km) away—three days by space ship. Even from this distance, the Moon's weak gravity pulls our oceans to create high and low tides. The Moon's surface is pitted with craters plus the footprints of 12 astronauts who have walked on it. There is no atmosphere to stop falling space rocks, and no weather to wash or wear the rocks or prints away.

Full Moon

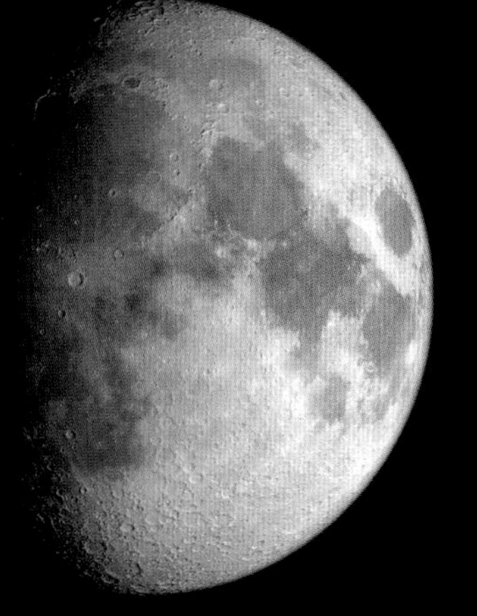

Waxing gibbous

Phases of the Moon

Half of the Moon is always lit by sunlight, but what can be seen from Earth changes in what we call "phases." Over a period of 29.5 days, the Moon goes through a cycle of phases from the New Moon, when the near side is in shadow, through the Full Moon, when the near side is fully illuminated, and back again to the New Moon.

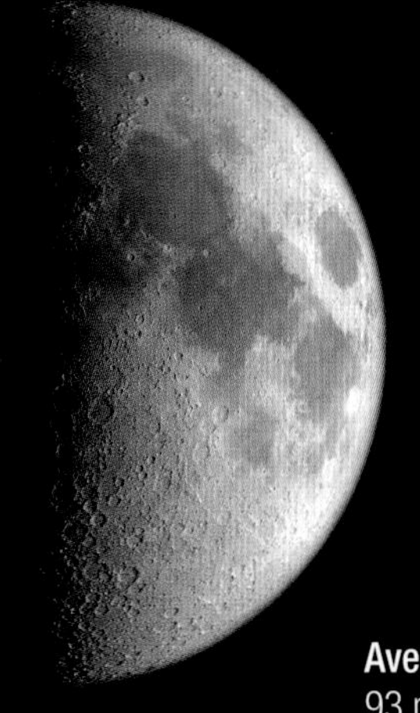

First quarter

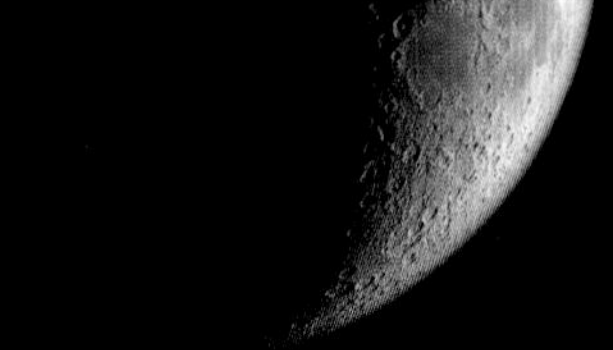

Waxing crescent

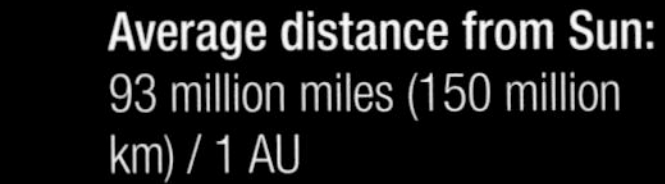

Average distance from Sun: 93 million miles (150 million km) / 1 AU

Average distance from Earth: 238,857 miles (384,403 km)

Speed of orbit round Earth: 2,288 mph (3,683 km/h)

Equatorial circumference: 6,783.5 miles (10,917 km)

Peak daytime temperature: 253°F (123°C)

Lowest nighttime temperature: –280°F (–173°C)

Atmosphere: small amounts of helium, argon, neon, ammonia, methane, carbon dioxide, as well as traces of sodium and potassium

Lunar lander Eagle descending to the surface

Astronaut Buzz Aldrin on the Moon

Eagle on the Moon

On July 21, 1969, Neil Armstrong and Buzz Aldrin made history as the first people ever to walk on the Moon. They had steered their fragile four-legged lunar module, Eagle, to a safe landing six hours earlier at 20.18 the previous day.

Worshipping the Moon

Moon goddesses and gods are important deities in many cultures, from ancient civilizations to the present day. In ancient Egypt, Thoth, for example, was god of the Moon, magic, and wisdom. Selene (left), goddess of the Moon, is depicted riding sidesaddle on a horse on this Greek coin struck in 1973.

7,000 deep **moonquakes** in **10 years** may be caused by **Earth's tidal pulls**.

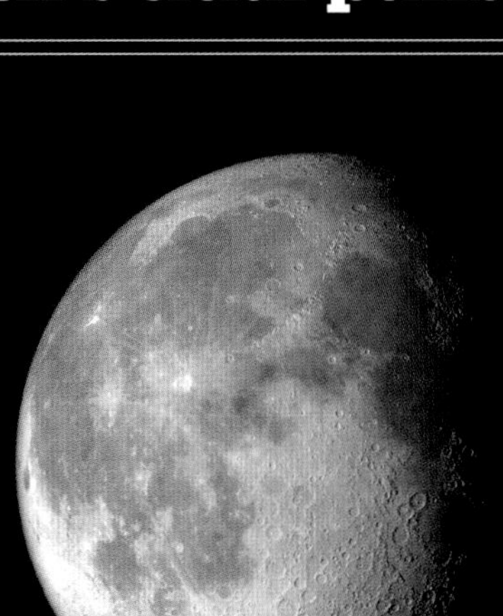

Waning gibbous

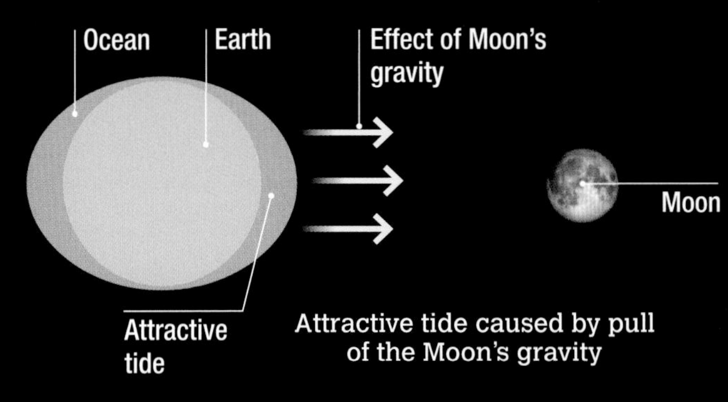

Attractive tide caused by pull of the Moon's gravity

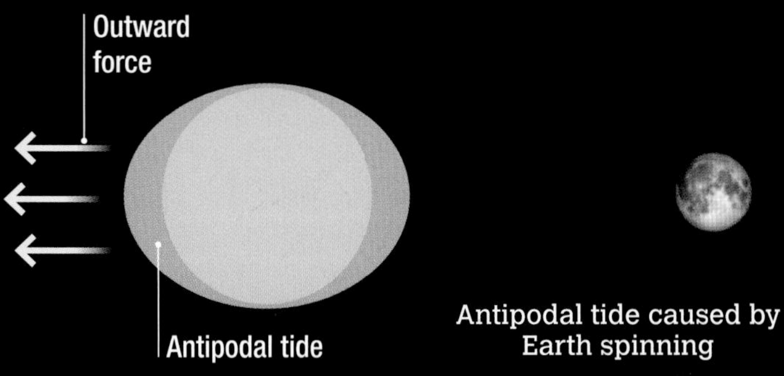

Antipodal tide caused by Earth spinning

Pull of gravity

Twice a day on Earth, sea levels rise and fall. These tides are mainly caused by the gravitational pull of the Moon, which lifts water into a bulge on the Earth's surface nearest to it. In an attractive tide, water is sucked toward the Moon and Sun. At the same time, on the opposite side of Earth, the outward force generated by the planet spinning pushes the water away in an antipodal tide.

Blood moon

About twice a year, when the Sun, Earth, and Moon line up and the Moon passes through Earth's shadow, the Moon appears a dull red or orange color. It has no light of its own, so when there is an eclipse like this, rays creep over the curves of Earth, and sunlight scattered through Earth's atmosphere bathes the surface of the Moon in red light.

First closeups

In 1609, Galileo (*see p.35*) drew a set of six watercolors recording phases of the Moon. He published these pictures the following year in *Sidereus Nuncius*, the first scientific book based on observations with a telescope.

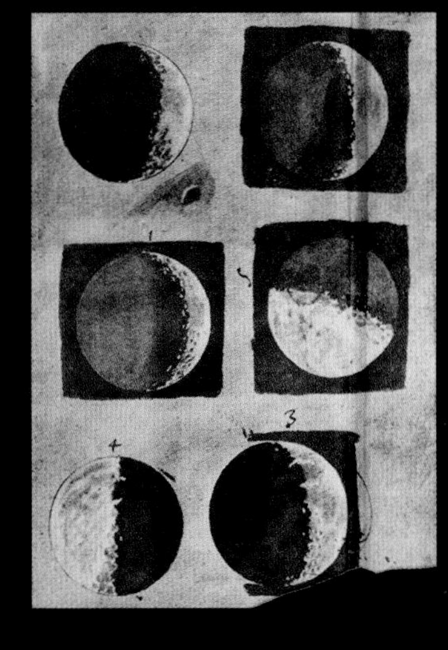

Last quarter

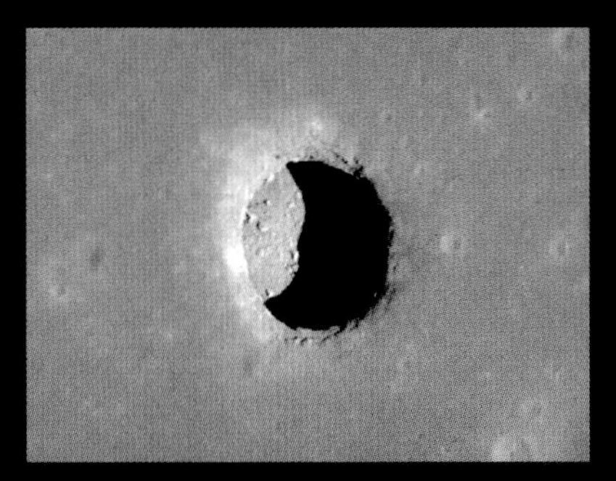

Moonbases in the future?

There are more than 200 lunar pits on the Moon ranging from 5–980 yds (5–984 m) across. Some of these have recently been identified as steep-walled pits that might lead to lava tubes. These caverns could in the future provide shelter for lunar bases.

Waning crescent

MAPPING THE MOON

NATURAL SATELLITE

The Moon's surface is dotted with craters blasted by comets and asteroids, as well as huge, smooth plains left by lava flowing from ancient volcanoes. There have been more than 60 missions to the Moon, and more than 40 spacecraft have landed on its surface, enabling astronomers to map it more accurately than any other body in space.

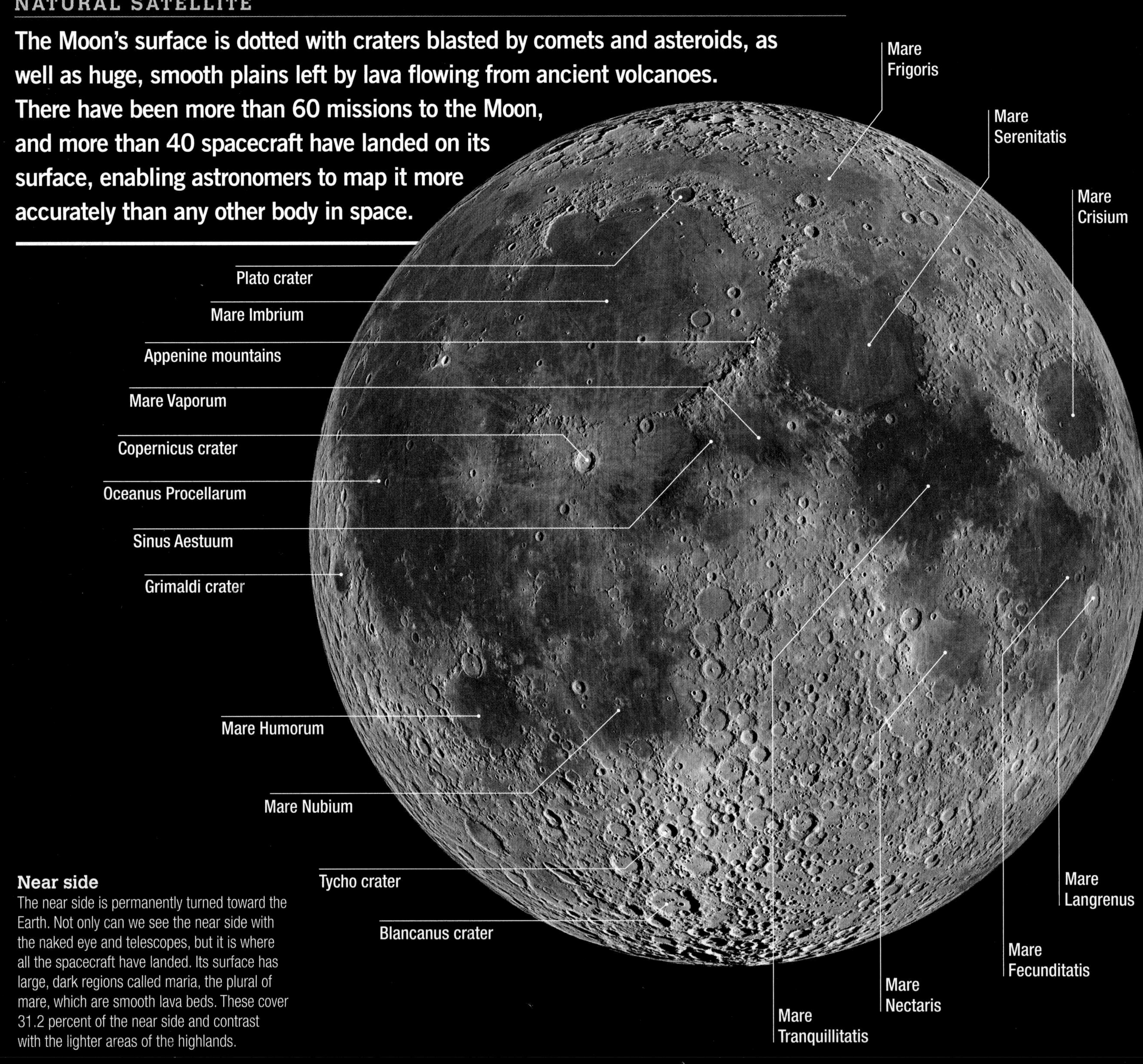

Near side

The near side is permanently turned toward the Earth. Not only can we see the near side with the naked eye and telescopes, but it is where all the spacecraft have landed. Its surface has large, dark regions called maria, the plural of mare, which are smooth lava beds. These cover 31.2 percent of the near side and contrast with the lighter areas of the highlands.

The **dark side** of the Moon is **a myth—both sides** of the Moon see the **same amount of sunlight** over a **lunar cycle**.

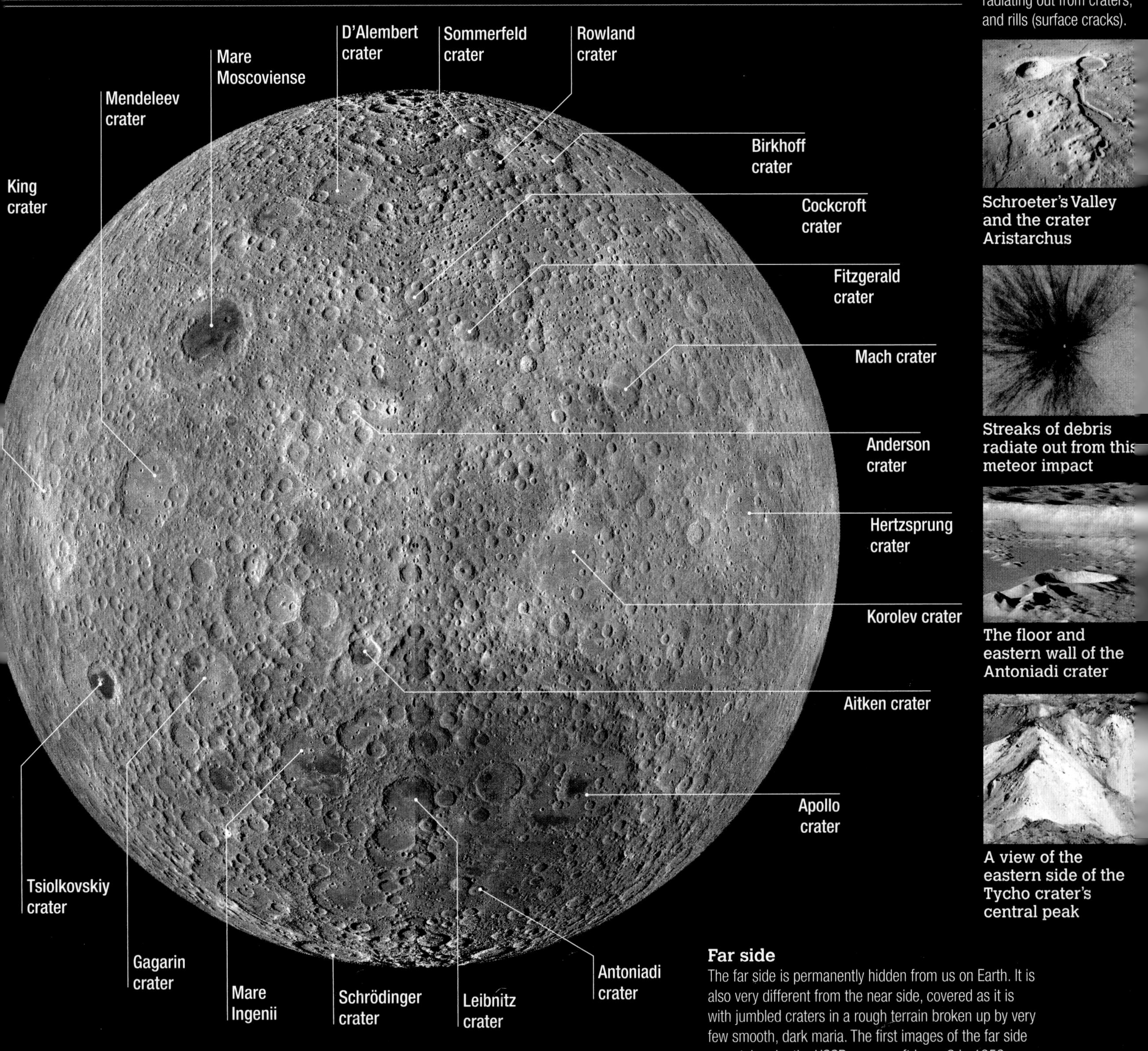

Far side

The far side is permanently hidden from us on Earth. It is also very different from the near side, covered as it is with jumbled craters in a rough terrain broken up by very few smooth, dark maria. The first images of the far side were taken by the USSR spacecraft Luna 3 in 1959.

Features of the Moon

As well as maria (plains) and terrae (highlands), both sides of the Moon have debris radiating out from craters, and rills (surface cracks).

Schroeter's Valley and the crater Aristarchus

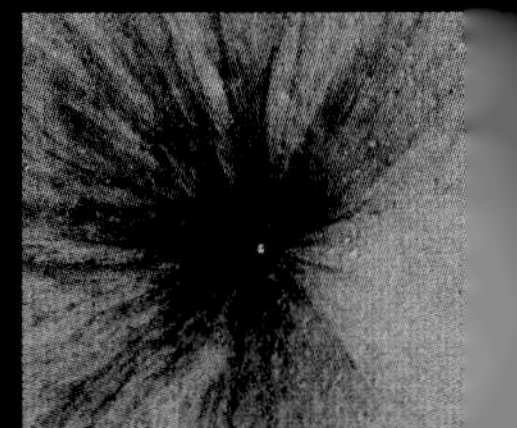

Streaks of debris radiate out from this meteor impact

The floor and eastern wall of the Antoniadi crater

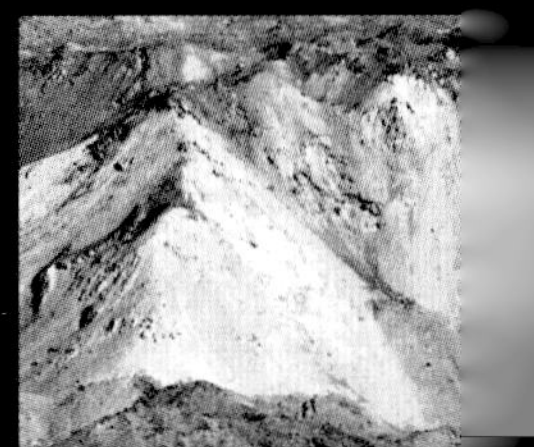

A view of the eastern side of the Tycho crater's central peak

Cover your eyes?

Eclipses have always drawn crowds, whether through fear or delight. This gathering of people took place in Times Square, New York, in 1932, with some people squinting through protective film or dark-coated glasses at a solar eclipse. Solar eclipses happen two to five times a year, always about two weeks before or after a lunar eclipse. Today, you can protect your eyes with special solar eclipse glasses, which are made of black polymer, a resin infused with carbon particles. These are usually 100,000 times darker than sunglasses and filter the light, blocking out all the ultraviolet rays, which can damage or destroy cells in the retina at the back of the eye.

ECLIPSES

Eclipses happen when a moon or planet moves into the shadow of another body in space. From Earth we can see two types of eclipse—of the Moon (lunar eclipse) and of the Sun (solar eclipse). For people today there is still something magical about that moment when darkness falls and the air chills because of these strange celestial dances, when birds fall silent and begin to roost, and nighttime animals emerge.

Transits

If a smaller body, such as a planet or the ISS, comes between a viewpoint and a larger body, such as the Sun, it is called a transit. Planet transits of Mercury and Venus, visible as black dots against the Sun, can be seen from Earth.

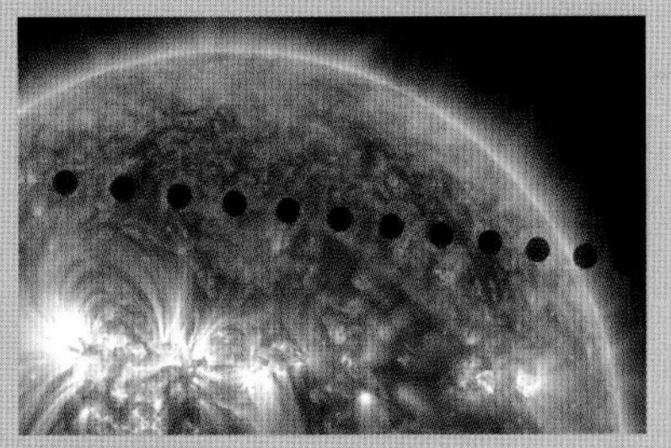

Composite transit of Venus

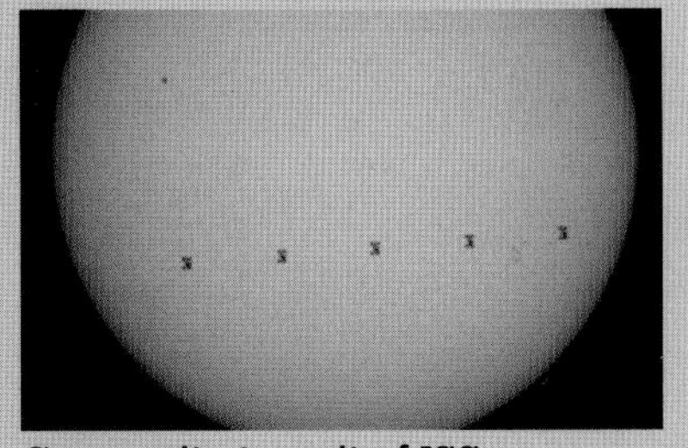

Composite transit of ISS

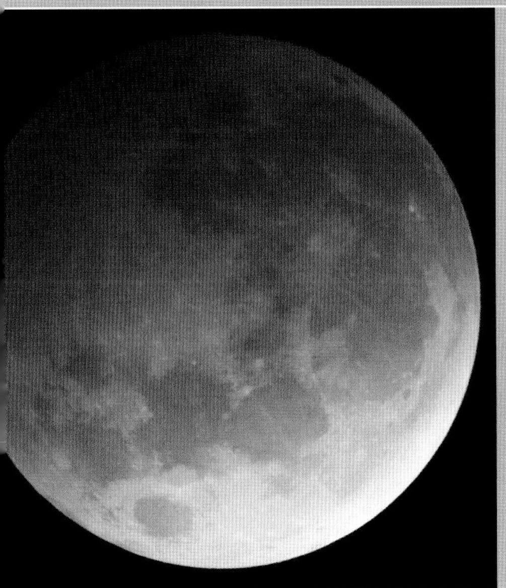

Total lunar eclipse

Lunar eclipses

A lunar eclipse happens when the Moon passes behind Earth and so is in its umbra, or shadow. There are three types of lunar eclipse. Total eclipses are the most dramatic because Earth blocks direct sunlight from reaching the Moon. During partial eclipses, Earth's shadow covers part of the Moon. During penumbral eclipses, the Moon is only subtly shaded.

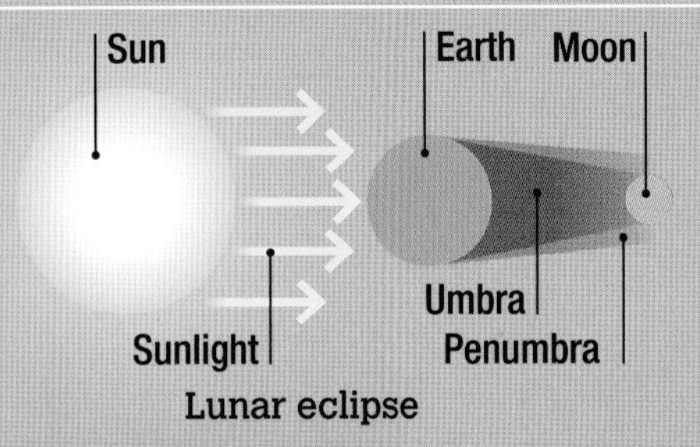

Lunar eclipse

Partial lunar eclipse

Solar eclipses

A solar eclipse happens when the Moon passes between the Sun and Earth, causing a shadow that falls on only part of Earth's surface. A total solar eclipse is when the Sun is covered completely by the Moon. In an annular eclipse, the Sun can be seen round the edges of the Moon. A hybrid eclipse will be seen as a total eclipse from certain points on Earth's surface, but viewed as an annular eclipse from other sites. In a partial eclipse, only part of the Sun is blocked by the Moon.

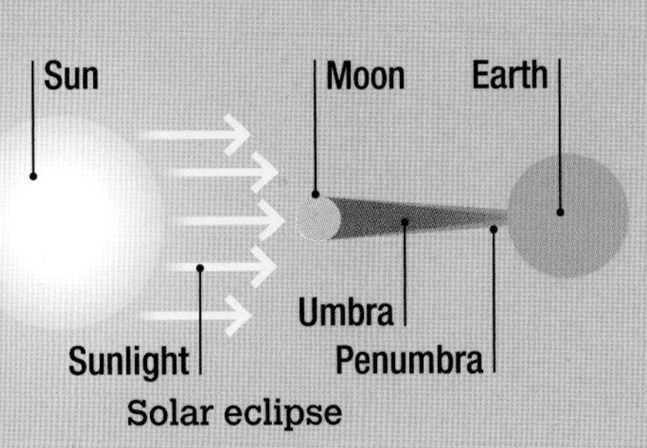

Solar eclipse

Annular eclipse

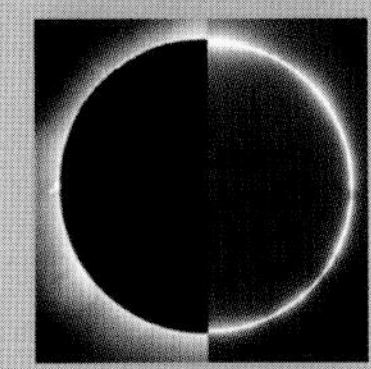

Hybrid eclipse

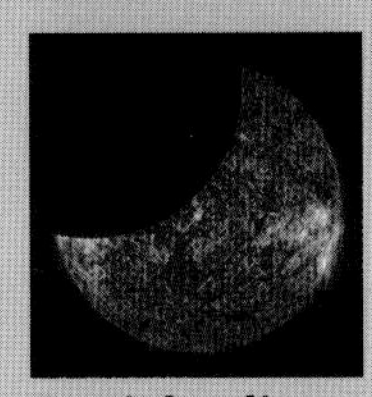

Partial eclipse

Total solar eclipse with Baily's Beads—sunlight shining through the Moon's uneven surface

Myths and eclipses

Throughout history, eclipses have had great significance. In China, Emperor Chung K'ang beheaded two astronomers because they put his realm in danger by failing to predict an eclipse. The Aztecs believed that demons of darkness would come and eat their people. In Thailand, it is said that Phra Rahu, the god of darkness, causes them by eating the Sun.

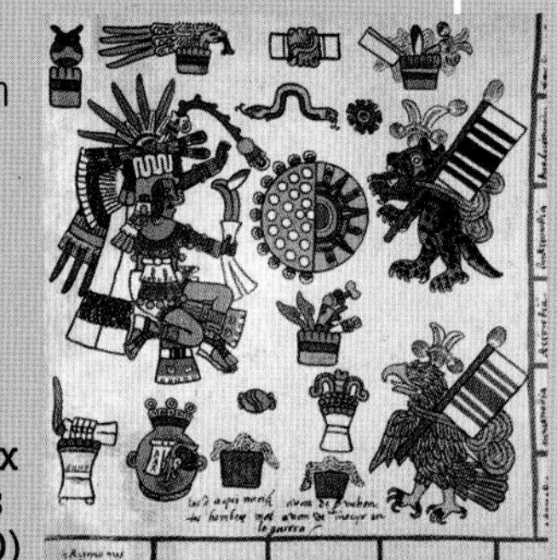

Eclipse (center) depicted in Aztec Codex Borbonicus (1519–1540)

The red planet

The ancient Egyptians called Mars "the red one," while to early Chinese astronomers it was "the fire star." Mars gets its rich red coloring from iron oxide in the rocks and soil. It has icy polar caps, but the thin atmosphere and cold temperatures mean that liquid water cannot exist on the surface.

Average distance from Sun: 142 million miles (228 million km) / 1.52 AU

First record: known to ancient civilizations

Main visits: Mariner 4 (1965), Mariner 9 (1971), Mars 3 (1971), Viking 1 (1976), Mars Pathfinder (1997), Spirit (2004), Opportunity (2004), Curiosity (2012)

Speed of orbit round the Sun: 53,858 mph (86,677 km/h)

Equatorial circumference: 13,259 miles (21,339 km)

Gravity: 38% of that on Earth

Year: 687 Earth days

Peak daytime temperature: 70°F (20°C)

Lowest nighttime temperature: –243°F (–153°C)

Atmosphere: carbon dioxide, nitrogen, argon

Known moons: 2

MARS

ROCKY PLANET

Mars holds some solar system records. It has the tallest volcano—Olympus Mons is three times the height of Everest; the longest canyon beyond Earth—Valles Marineris stretches 2,500 miles (4,000 km) or the breadth of the USA; and the biggest dust storms, with huge clouds that surround the whole planet for months.

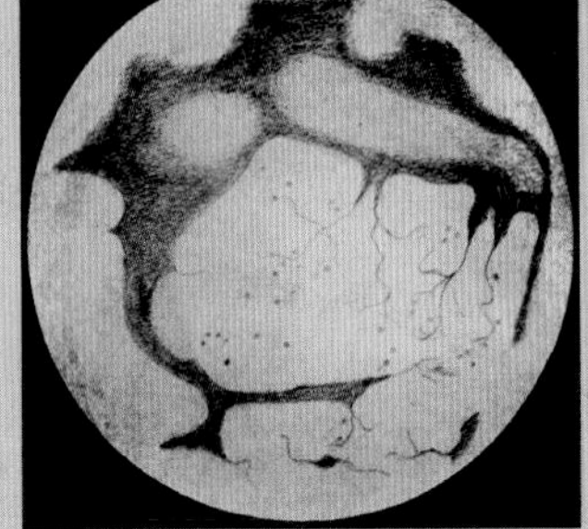

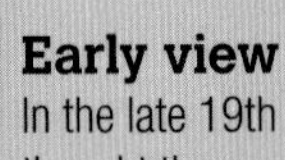

Early view

In the late 19th century, many people thought there were channels or canals for flowing water visible on Mars. In 1902 and 1903, British astronomer Edward Maunder published the results of an experiment that he had done to show that the "canals" were an optical illusion.

This **cold, dry desert world** is **half** the diameter of **Earth**, but has a similar **landmass**.

Phobos

Deimos

Two moons

In 1877, US astronomer Asaph Hall, urged on by his wife Angeline Stickney, discovered two moons traveling around Mars. The inner moon, the brightest, he named Phobos, meaning "fear," while he chose Deimos, "panic," for the potato-shaped outer moon.

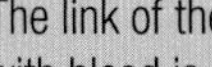

War god

The link of the red planet with blood is inescapable. The Romans named it Mars, for their god of war. The ancient Babylonians called the planet Nergal, for their god of fire and war.

Blocks of deep-basin deposit

Estimated extent of ancient Martian sea

Life on Mars?

NASA's Mars Reconnaissance Orbiter (MRO) has found 3.7 billion-year-old hydrothermal deposits in the Eridania basin, a southern region. These beds once held ten times as much water as all of the Great Lakes combined and may have provided the environment necessary for life on Mars.

Earthbound

Before space probes reached Mars, scientists relied on rare meteorite fragments that reached Earth to discover more about the planet. Of the 61,000 meteorites discovered on our planet, only 131 have been identified as Martian.

Features of Mars

Great images have been sent back from Mars by the many probes that have visited the planet, showing a wide variety of landforms.

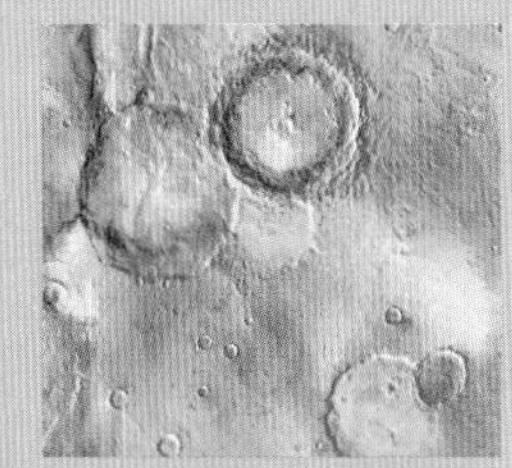

Ancient lake beds

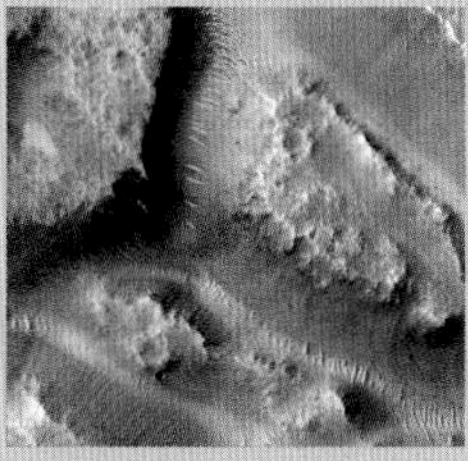

Valley network

Sedimentary rock formations

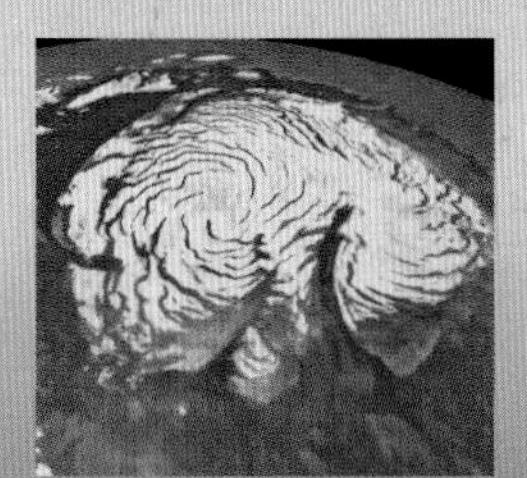

Polar ice caps

Dusty atmosphere

Mars is covered in fine, dry particles of red dust. The planet's thin atmosphere frequently creates winds that whip these particles into massive dust storms, some of which cover the entire planet.

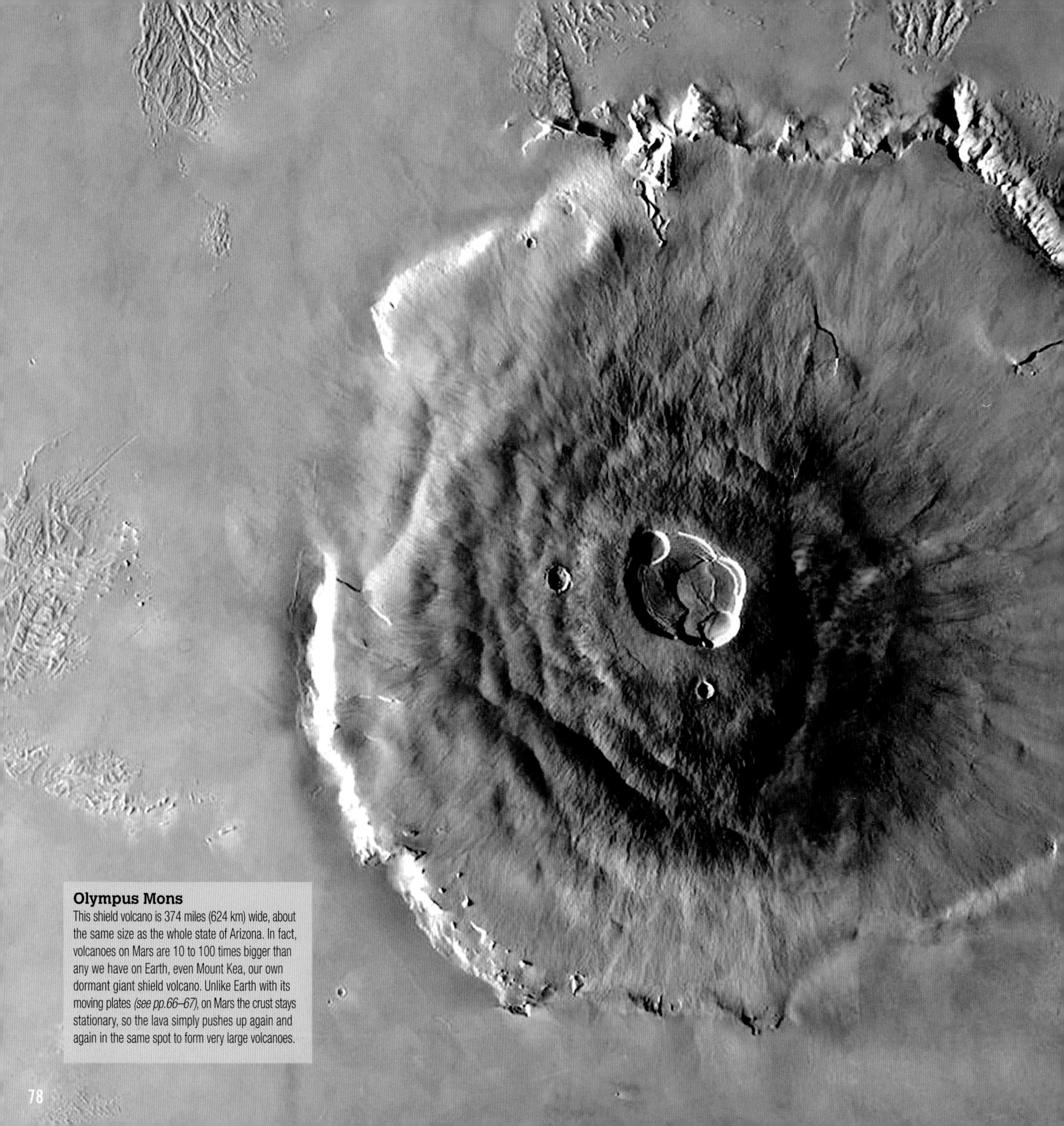

Olympus Mons
This shield volcano is 374 miles (624 km) wide, about the same size as the whole state of Arizona. In fact, volcanoes on Mars are 10 to 100 times bigger than any we have on Earth, even Mount Kea, our own dormant giant shield volcano. Unlike Earth with its moving plates *(see pp.66–67)*, on Mars the crust stays stationary, so the lava simply pushes up again and again in the same spot to form very large volcanoes.

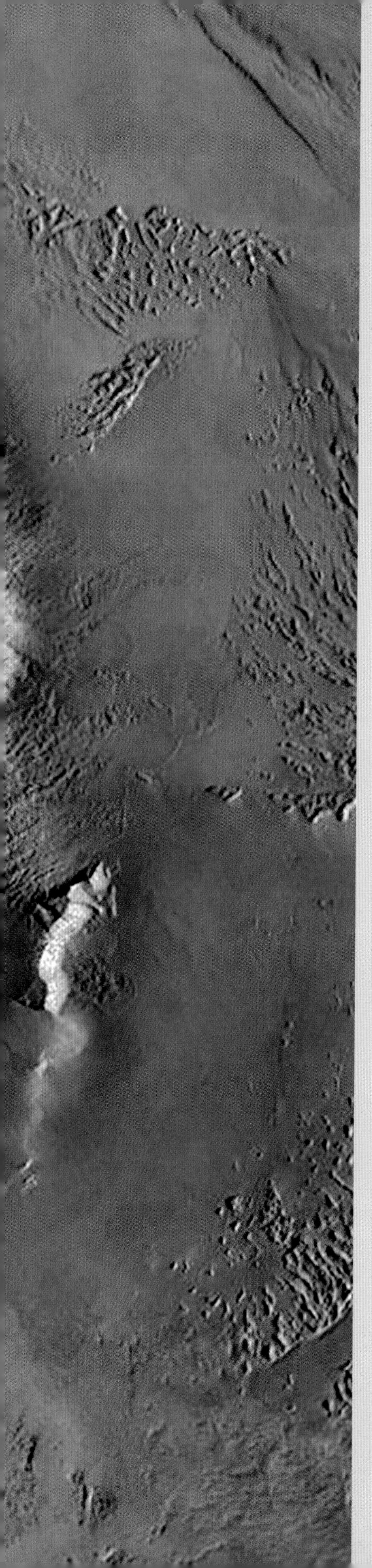

VOLCANOES IN THE SOLAR SYSTEM

BY DR. ROSALY LOPES, SENIOR RESEARCH SCIENTIST, NASA'S JET PROPULSION LABORATORY, CALTECH

Earth is not the only place that has volcanoes—there are some monster mountains out in space spewing out melted rock or frozen gases. This happens when heat and energy build up in a planet or moon's core, and then burst out, with spectacular results.

Earth is shaped by the existence of volcanoes—the islands of Hawaii, for example, formed from the terrifying force of molten rock that was blasted up from the deep furnaces under the north Pacific Ocean more than 2.5 million years ago.

Early in the life of our solar system, the planets and moons had higher internal temperatures and so there were many active volcanoes. Evidence of this is seen on our Moon's surface in the huge, dark plains of solid lava that once flowed from them, as well as in lunar domes or hummocks, formed by slowly cooling magma.

Mars has the largest volcano in the solar system. Olympus Mons, at a height of 16 miles (25 km), is nearly three times taller than Mount Everest. This enormous volcano was one of many that gushed fields of lava over the planet's surface. Mercury had active volcanoes, and scientists believe that Venus still has many, hidden under dense clouds.

Volcanologist
Dr. Rosaly Lopes is the discoverer of the most active volcanoes anywhere, having found 71 on the moon Io alone. She is currently studying Saturn's moon Titan and its strange ice volcanoes.

Today, with the solar system slowly cooling, all the active volcanoes apart from those on Earth are on remote moons. Most dramatic of these are on Io, one of Jupiter's moons and the most volcanically active body in our solar system. The powerful gravity of Jupiter shifts the insides of this moon so much that the friction generates enough heat to melt rock and gases, pushing hot lava hundreds of miles out into its atmosphere.

But not all volcanoes blast out molten rock and ash. Some are called cryovolcanoes because they erupt cold or frozen gases such as water, ammonia, or methane. This happens when pressurized water sits just below the outer surface of the volcano. When heat from the core builds up, the volcano fires out jetlike plumes of vapor and ice particles. Great examples of this are the 101 geysers on Saturn's moon Enceladus (*see p.96*). These blast water and chemicals far out into space.

"One of **Io's** volcanoes, **Loki**, is more **powerful** than **all** of **Earth's volcanoes** combined."

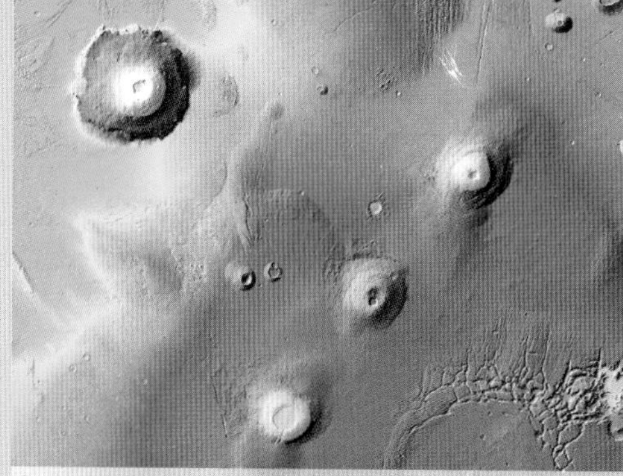

Four of Mars' major volcanoes

Active Mauna Loa, Earth's largest volcano

Eruption on Jupiter's moon Io

Volcano Sapas Mons on Venus

Rough surface

In 1997, NASA's Pathfinder mission to Mars sent back pictures of a rock-covered region of Ares Vallis, north of the planet's equator, and this is one of the composite images. The "Twin Peaks" hills, some 100 ft (35 m) tall, are seen in the background. The North Twin is around 2,800 ft (860 m) away from the Sojourner rover, and the South Twin is about 3,300 ft (1,000 m) away. On this mission, Sojourner spent 83 days exploring the Martian surface and took chemical, atmospheric, and other measurements.

THE MARS ROVERS

PLANET EXPLORATION

Is there life on Mars? Robot geologists tell us there could have been once. A series of missions have sent machines to test the geology, environment, and atmosphere of Mars. Their equipment can sniff the air, drill into rocks, and blast boulders with lasers. They have taught us that the sunset on Mars is blue, and shown us that water once flowed and tiny microbes could have lived there.

Family portrait
This group of two test rovers and a flight spare shows the three generations of Mars rovers developed at NASA's Jet Propulsion Laboratory. The tiny Sojourner spare stands in front of a working test version of Spirit and Opportunity. On the right is a Mars Science Laboratory test rover the size of Curiosity.

Landing sites
There is only one chance to land at a particular site on a distant planet, so it is critical that the scientists get it right. NASA holds landing site workshops and invites scientists from all over the world to take part in making the decision.

Camera
The rover had two forward black-and-white cameras and one rear color camera.

Mobile mass: 23 lb (10.6 kg)

Launch date: December 4, 1996

Rocket: Delta II 7925

Launch site: Cape Canaveral Air Force Station, Florida

Landing site: Ares Vallis

Deployed from: Mars Pathfinder

Deployment date: July 4, 1997

Status: Last contact September 27, 1997

Sojourner
This was the first time that a wheeled robot (it is the size of a microwave oven) had studied the surface of another planet. The rover took images, analyzed rocks and soil, and studied the weather.

Color image of tire tracks

Sojourner image of laser track

Tire tracks and a dust devil

Columbia Hills inside the Gusev crater

Rocks of volcanic origin

Mobile mass: 397 lb (180 kg)

Launch date: Spirit: June 10, 2003; Opportunity: July 8, 2003

Rocket: Spirit: Delta II 7925; Opportunity: Delta II 7925H

Launch site: Cape Canaveral Air Force Station, Florida

Landed: Spirit: January 3, 2004; Opportunity: January 25, 2004

Landing sites: Spirit: Gusev Crater; Opportunity: Meridiani Planum

Status: Spirit: end of mission March 22, 2010; Opportunity: still operating

Spirit and Opportunity
Both rovers found evidence of past water, as well as volcanic activity. They detected argon in the atmosphere and showed how the wind creates dust devils.

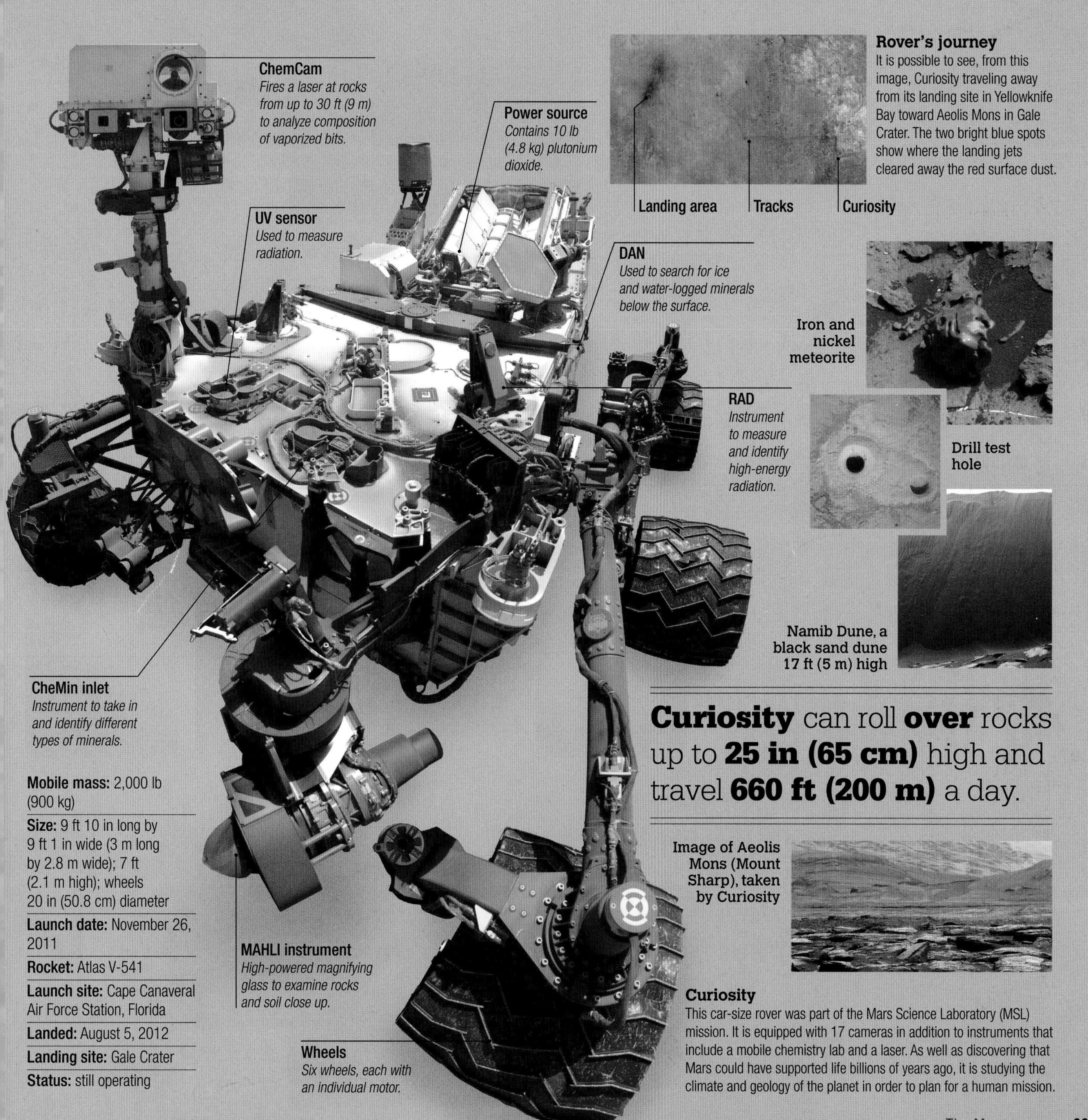
ChemCam
Fires a laser at rocks from up to 30 ft (9 m) to analyze composition of vaporized bits.
Power source
Contains 10 lb (4.8 kg) plutonium dioxide.
UV sensor
Used to measure radiation.
DAN
Used to search for ice and water-logged minerals below the surface.
RAD
Instrument to measure and identify high-energy radiation.
CheMin inlet
Instrument to take in and identify different types of minerals.
MAHLI instrument
High-powered magnifying glass to examine rocks and soil close up.
Wheels
Six wheels, each with an individual motor.
Rover's journey
It is possible to see, from this image, Curiosity traveling away from its landing site in Yellowknife Bay toward Aeolis Mons in Gale Crater. The two bright blue spots show where the landing jets cleared away the red surface dust.
Landing area
Tracks
Curiosity
Iron and nickel meteorite
Drill test hole
Namib Dune, a black sand dune 17 ft (5 m) high
Curiosity can roll over rocks up to 25 in (65 cm) high and travel 660 ft (200 m) a day.
Image of Aeolis Mons (Mount Sharp), taken by Curiosity
Curiosity
This car-size rover was part of the Mars Science Laboratory (MSL) mission. It is equipped with 17 cameras in addition to instruments that include a mobile chemistry lab and a laser. As well as discovering that Mars could have supported life billions of years ago, it is studying the climate and geology of the planet in order to plan for a human mission.
Mobile mass: 2,000 lb (900 kg)
Size: 9 ft 10 in long by 9 ft 1 in wide (3 m long by 2.8 m wide); 7 ft (2.1 m high); wheels 20 in (50.8 cm) diameter
Launch date: November 26, 2011
Rocket: Atlas V-541
Launch site: Cape Canaveral Air Force Station, Florida
Landed: August 5, 2012
Landing site: Gale Crater
Status: still operating

ASTEROID BELT

ORBITING SPACE ROCKS

Between the orbits of Mars and Jupiter, millions of lumps of rock zoom through space along the Asteroid Belt. These leftovers from the creation of our solar system range in size from small boulders to sizeable masses big enough to be called dwarf planets—the biggest, Ceres, is 592 miles (953 km) across. Some contain valuable minerals that we may one day harvest. They occasionally hit each other, and—very rarely—spin out of orbit and strike planets, including Earth.

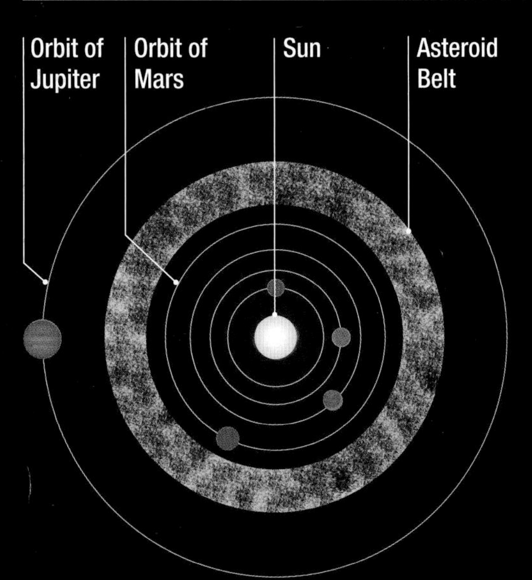

Finding the Belt

On January 1, 1801, Giuseppe Piazzi, an Italian priest and astronomer, discovered and named the dwarf planet Ceres. Fifteen months later, Pallas was discovered. By 1850, astronomers were talking about the "Asteroid Belt."

Asteroid alert

NASA's NEOWISE spacecraft uses infrared to identify and track asteroids. This edge-on view of our solar system shows asteroids NEOWISE has identified, including 28 Near Earth Objects (NEOs). The largest so far is Florence, which came close to Earth in September 2017.

Orbit of Earth

Sun

Asteroids in the Belt

It is estimated that there are only up to 1.9 million of these chunks of rock and metal in the Belt that are larger than 0.6 miles (1 km) in diameter, compared to the millions of smaller asteroids, some only the size of pebbles. Five of the largest asteroids are shown here (comparative sizes in diameters), as well as three of the smaller ones.

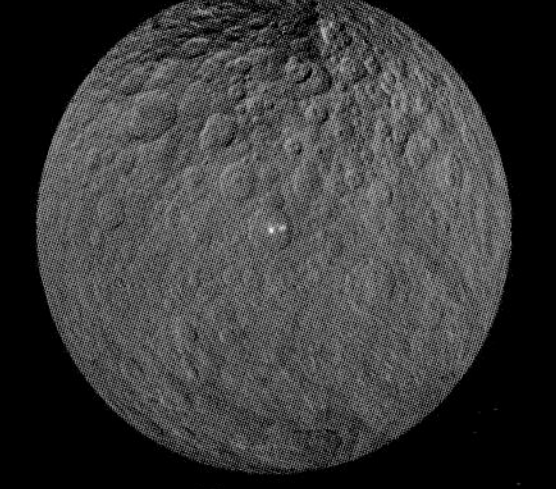
Ceres
592 miles (953 km)

Vesta
329 miles (530 km)

Pallas
318 miles (512 km)

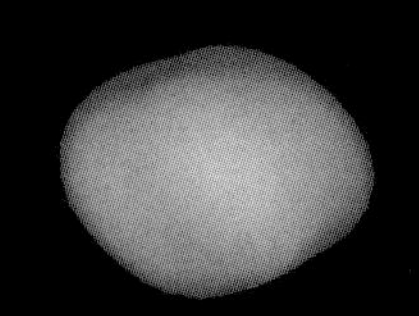
Europa
195 miles (315 km)

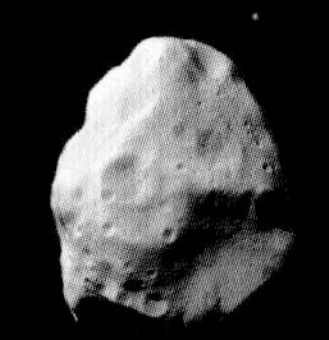
Lutetia
75 miles (121 km)

The **Asteroid Belt** contains many **millions** of asteroids, **most** of them less than **33 ft (10 m)** across.

Orange dots are potentially hazardous asteroids (PHAs)

Blue dots are near-Earth asteroids (NEAs)

Ida (and its moon Dactyl)
19.5 miles (31.4 km)

Eros
10.4 miles (16.84 km)

Itokawa
1,150 ft (350 m)

Asteroid mission

The Dawn probe was launched in September 2007, its mission to find out everything it can about two of the largest, but very different, asteroids. Dawn orbited rocky Vesta in 2011–2012, and in 2015 reached icy Ceres, which it is still orbiting. The probe has found evidence that Vesta is the source of many meteorites, and that there is organic material on Ceres.

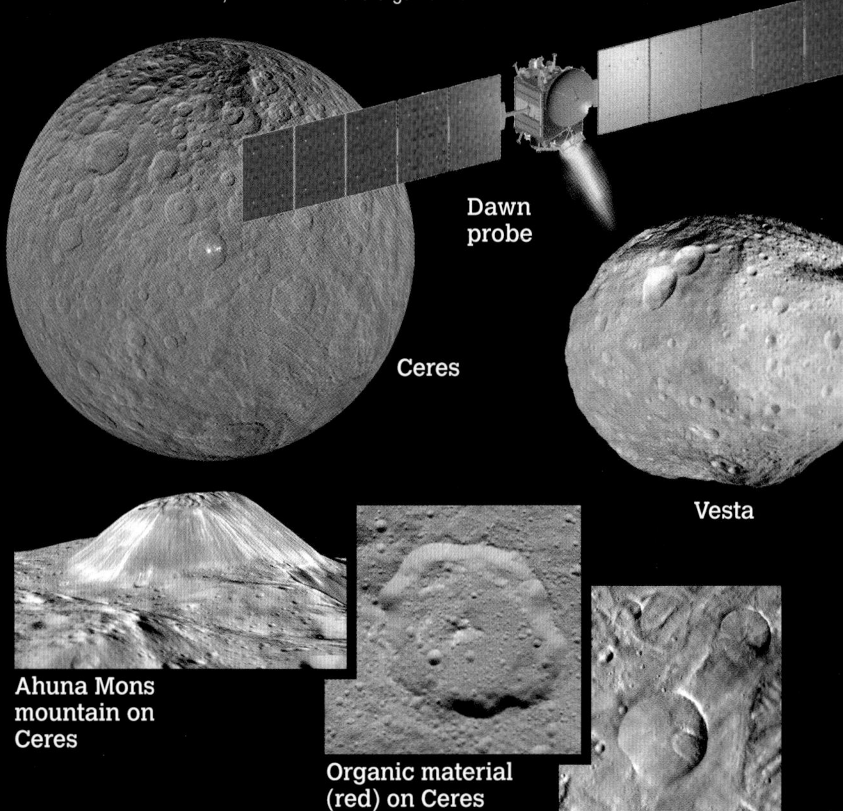

Dawn probe

Ceres

Vesta

Ahuna Mons mountain on Ceres

Organic material (red) on Ceres

Fresh impact crater on Vesta

Threats to Earth

About once a year, a car-sized asteroid hits Earth's atmosphere, creating a fireball and burning up. And every 2,000 years, a meteoroid the size of a soccer pitch hits Earth and causes great damage. But once every few million years, an object large enough to wipe out life on Earth threatens the planet.

Up to 82 ft (25 m) diameter—burns up in the atmosphere

165-ft (50-m) diameter—creates a crater 0.7 miles (1.16 km) wide

9-mile (15-km) diameter—wiped out the dinosaurs on Earth 65 million years ago

JUPITER

GAS GIANT

All the other planets in our solar system would fit inside Jupiter, but they would never settle. This planet is a giant ball of swirling, stinking gases and liquids, mostly hydrogen and helium, swept around in the jet streams created by its ten-hour rotation. At its equator, the clouds move at more than 28,000 mph (45,000 km/h)—fast enough to make the planet bulge slightly.

Average distance from Sun: 484 million miles (778 million km) / 5.2 AU

First record: known to ancient civilizations

Main visits: Pioneers 10 and 11 (1973), Voyager 1 (1977), Voyager 2 (1979), Galileo (1995), Cassini-Huygens (2000), New Horizons (2007), Juno (2016)

Speed of orbit round the Sun: 29,205 mph (47,002 km/h)

Equatorial circumference: 279,118 miles (449,197 km)

Gravity: 2.528% of that on Earth

Year: 4,333 Earth days

Temperature: –148°C (–234°F)

Atmosphere: hydrogen, helium, methane, water, ammonia,

Known moons: 53 confirmed, 16 provisional

Rings: 4

The **Great Red Spot** is actually **shrinking**—it is now **10,160 miles** (16,350 km) **across**.

Great Red Spot

A red spot, seen by astronomers on Earth for centuries, is twice the width of Earth and marks a violent storm where wind speeds reach 400 mph (645 km/h). This anticyclone in the planet's upper atmosphere is trapped between two jet streams, and swirls around a center of high atmospheric pressure that makes it rotate.

Makeup of a giant

This planet is so large that more than 1,300 Earths could fit inside it. Surrounded by thick, colorful clouds of poisonous gas, it has at its center a rocky core that is slightly bigger than Earth. Around the core is an ocean of liquid hydrogen about 620 miles (1,000 km) deep.

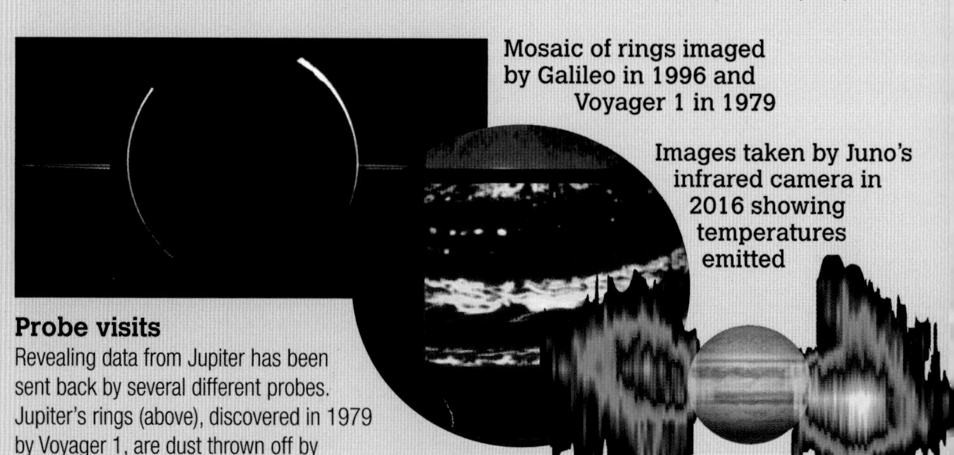

Mosaic of rings imaged by Galileo in 1996 and Voyager 1 in 1979

Images taken by Juno's infrared camera in 2016 showing temperatures emitted

Radio emission measurements taken by Cassini in 2001

Probe visits

Revealing data from Jupiter has been sent back by several different probes. Jupiter's rings (above), discovered in 1979 by Voyager 1, are dust thrown off by impacts on small moons. There are two faint outer rings which are called "gossamer rings," a wide, flat main ring, and a thick inner ring called a "halo."

Features of Jupiter

From Earth, Jupiter appears to be made up of "stripes" of dark belts and light zones. Probes have tracked their movement and revealed that they are bands of gas that remain separate from one another despite their stormy nature.

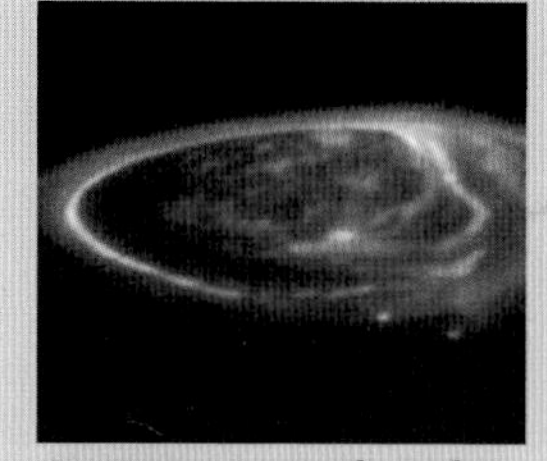

Aurora at north pole

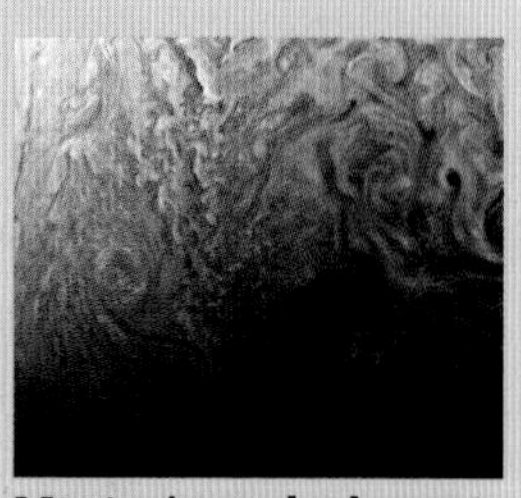

Mysterious dark spot among the storms

Huge counter-clockwise storm

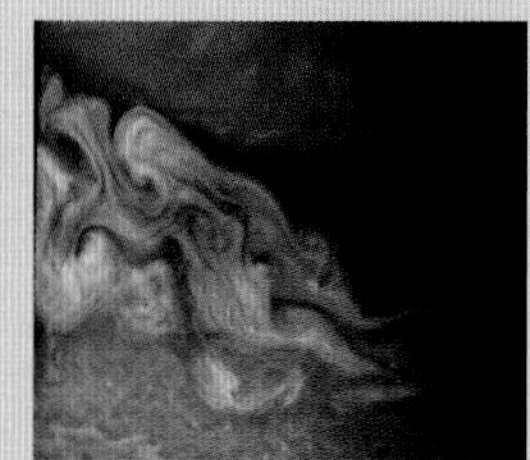

Wave formation of clouds

Lightning storms
The white spots are lightning-bearing thunderstorms.

Hot spots
All around the northern edge of the equitorial zone are dark patches where there are relatively few clouds and heat can escape.

White clouds
These high clouds consist of frozen ammonia crystals.

Zone interface
Differences in color between zones are due to the amount of trace chemicals in each layer.

Stormy chaos
The storms whirling round the planet are at the heart of a turbulent chaotic system.

Northern hemisphere

Jupiter spins round very fast and this gives rise to very strong weather patterns. The layers we see in the clouds that surround this giant planet are caused by alternating eastward and westward jet streams. The north pole is at the top of this picture, obscured by an atmospheric haze.

North pole
This is less clearly visible because it is viewed at an angle by Cassini through thicker atmospheric haze.

Different shapes
Many clouds appear in streaks because they are stretched by the turbulence.

In perspective
The smallest visible features of these images taken by Cassini are about 75 miles (120 km) across.

Great Red Spot
This storm rotates counter-clockwise over a period of 14 Jovian days (6 Earth days).

Oval BA
This red storm in the southern hemisphere is similar to, but smaller than, the Great Red Spot.

Whirlwinds
Continual stretching and folding by the winds cause the wave formations.

South pole
Like the north pole, this is imaged through a thick atmospheric haze.

Southern hemisphere

Jupiter is the stormiest planet in our solar system, and the Great Red Spot, the largest storm, dominates the southern hemisphere. The cloudy sphere's bright belts change their shape all the time as the 10-hour rotation of the planet ensures that they are in constant motion.

GALILEAN MOONS

JUPITER'S SATELLITES

With four large moons and many smaller satellites, Jupiter forms a kind of miniature solar system. The "big four" were discovered by Italian astronomer Galileo in 1610. At first he thought they were stars, but then he realized they actually orbited Jupiter. His observations supported Copernicus' theory that the Sun, not the Earth, was the center of our solar system.

Discovery
On a cold, clear evening in January 1610, Galileo Galilei pointed his homemade telescope toward Jupiter. He saw two "stars" to the east and one to the west. The following evening he realized that the stars had moved, and after observing them for several days, he established the existence of four moons.

EUROPA

ICY WORLD

Europa is a frozen moon with the smoothest surface of any object in our solar system. Some scientists believe that 60 miles (100 km) underneath its icy outer shell is a vast salty ocean holding twice as much liquid water as Earth.

Distance from the Sun: 484 million miles (778 million km)

Orbital distance from Jupiter: 414,000 miles (670,900 km)

Orbit time: 3.6 days

Diameter: 1,900 miles (3,100km)

Age of surface: 20–180 million years

Surface temperature: never above –260°F (–160°C)

Discovered by: Galileo Galilei in January 1610

Atmosphere: oxygen, too thin to breathe

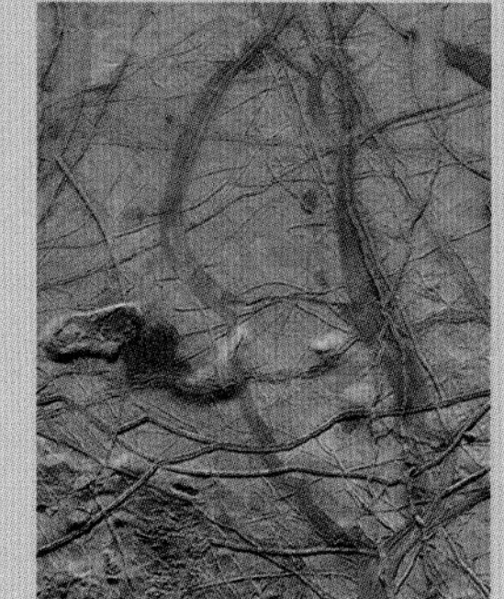

Evidence of water
In 2016, the Galileo spacecraft took pictures of these cracks that cover the icy surface of Europa. Newer fractures cut across older ones, and dark bands are visible where the surface has re-frozen in the past. It is thought that the tides of the ocean beneath cause the cracks.

Icy moon
In this picture, put together by images taken by the Galileo probe in the late 1990s, areas that are blue or white contain almost pure water ice. This moon's icy surface gives a high degree of "albedo," or light reflectivity.

CALLISTO

BRIGHT MOON

This moon lies a million miles from Jupiter, well out of range of the gravity that stretches and pulls the other moons. It is pitted with thousands of craters, the largest of which is 190 miles (305km) across.

Orbital distance from Jupiter: 1,168,000 miles (1,880,000 km)

Orbit of Jupiter: 16.7 days

Diameter: 2,985 miles (4,800 km)

Age of surface: about 4 billion years

Surface temperature: average –218.47°F (–139.2°C)

Discovered by: Galileo Galilei in January 1610

Atmosphere: thin layer of carbon dioxide

Lit up
Jupiter's second largest moon is composed of rock and ice in equal measures. A thick surface of ice means that Callisto reflects a higher proportion of sunlight back into space compared to our own Moon.

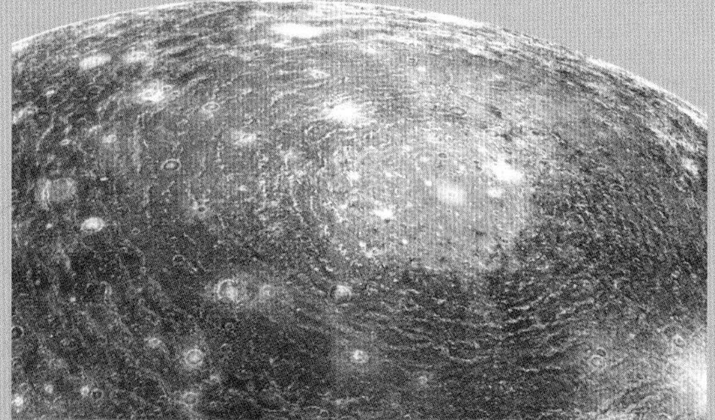

Cracked surface
Callisto's surface is covered with impact craters. There are no volcanoes or even any large mountains. It is one enormous ice-field, covered with cracks and craters that are the result of billions of years of collision with interplanetary debris.

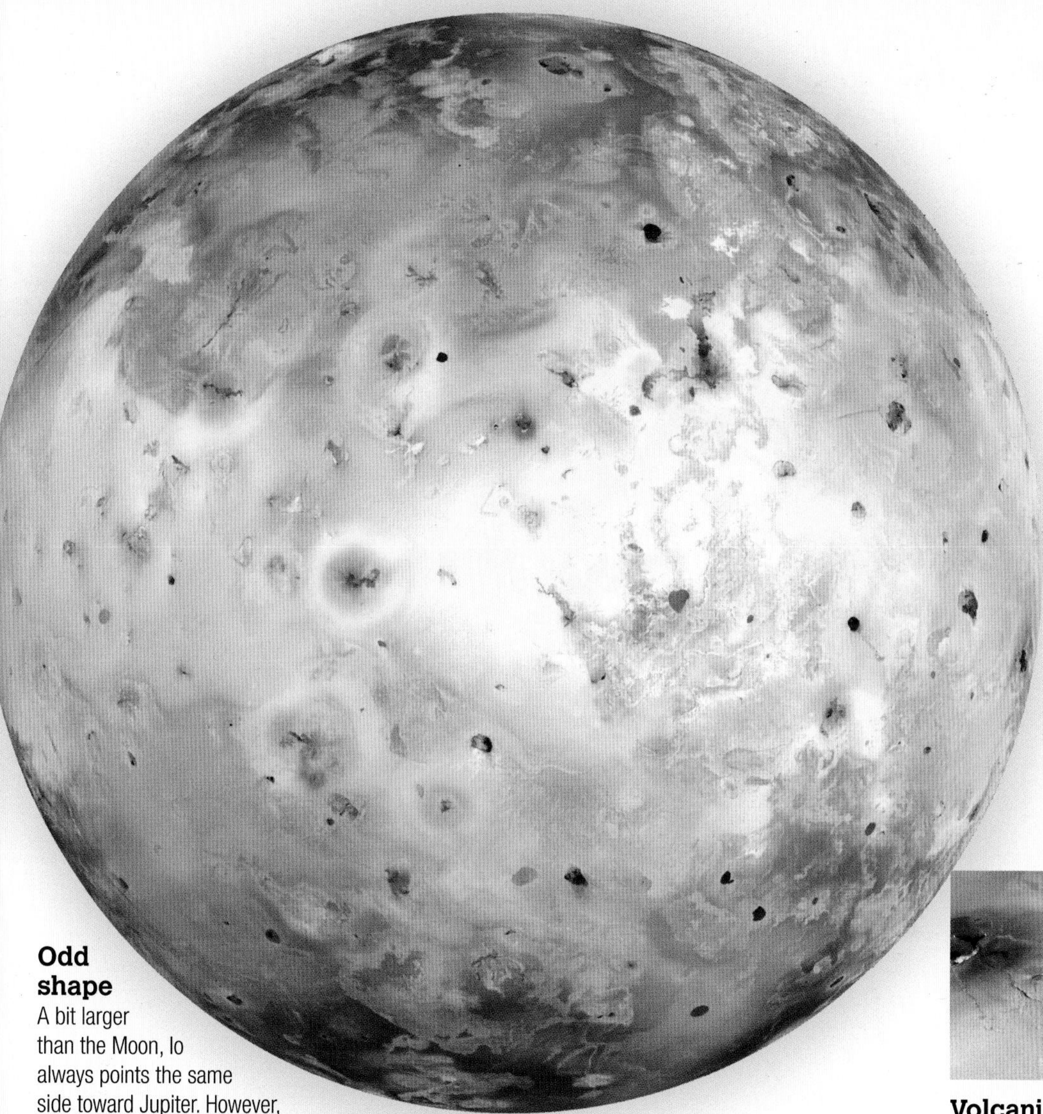

IO

ACTIVE WORLD

The innermost moon, Io, sits above Jupiter in a plasma cloud, which is held there by the planet's strong magnetic field. Io is a violent, dangerous place studded with hundreds of volcanoes that blast molten sulfur up to 190 miles (300 km) into the atmosphere. At times they are so powerful they are seen with large telescopes on Earth.

Odd shape
A bit larger than the Moon, Io always points the same side toward Jupiter. However, the pulls of the orbits of Europa and Ganymede cause tremendous tidal forces that make Io's surface bulge by up to 330 ft (100 m).

Infrared image showing volcano Loki Patera

Eruption on Io's surface

Volcanic action
Io is the most volcanically active world in our solar system. Despite its ice-cold sulfur dioxide snowfields, the heat near volcanoes can reach 3,000°F (1650°C).

Distance from the Sun: 484 million miles (778 million km)

Orbital distance from Jupiter: 262,000 miles (422,000 km)

Orbit of Jupiter: 1.77 days

Diameter: 2,264 miles (3,643 km)

Age of surface: about 4 billion years

Surface temperature: average –202°F (–130°C)

Discovered by: Galileo Galilei in January 1610

Atmosphere: a thin layer of sulfur dioxide

GANYMEDE

MAGNETIC MOON

Giant Ganymede would be classed as a planet if it orbited the Sun instead of Jupiter, and is the largest moon in our solar system. Layers of silicate rock and water ice rest on a core of metallic iron and iron sulfide that make this the only moon to generate a magnetic field.

Distance from the Sun: 484 million miles (778 million km)

Orbital distance from Jupiter: 665,000 miles (1,070,000 km)

Orbit of Jupiter: 7.1 days

Diameter: 3,270 miles (5,262 km)

Age of surface: about 4 billion years

Surface temperature: average –218.47°F (–139.2°C)

Discovered by: Galileo Galilei in January 1610

Atmosphere: thin layer of oxygen

Size of a planet
This moon is larger than the planet Mercury. It has a metallic iron core surrounded by rock, and an icy shell for a surface. The highly cratered dark terrain in this image is believed to be the original crust of the moon. The bright spots are large craters—Tros (upper right) and Cisti (lower left).

In the night sky

Some comets are seen only once in our night sky and are known as non-periodic comets. The Great Comet of 2007, McNaught—seen here over the Lower Eyre Peninsula in southern Australia—was discovered on August 7, 2006, by British-Australian astronomer Robert H. McNaught. It was the brightest comet for 40 years.

COMETS

Early sky-watchers called fiery-tailed comets "long-haired stars." These dirty snowballs of ice, rock, dust, and gas orbit the Sun, so we see them more than once. The famous Halley's comet was first recorded in 240 BCE and visits every 75.3 years. Comets also leave trails of debris that can lead to meteor showers on Earth.

A long chase

On March 2, 2004, ESA's Rosetta probe launched and rocketed out into the solar system. Ten years later, in 2014, it arrived at its target, Comet 67P/Churyumov-Gerasimenko, and deployed its lander, Philae. But Philae, which was solar-powered, landed on its side in an area that sunlight could not reach. Rosetta sent back valuable information but was deliberately crashed into the comet in September 2016.

Rosetta probe

Streaming tails

As they zoom through space near the Sun, comets heat up and form two tails, one of dust and the other of plasma, or ionized gas. The dust tail appears white, while the plasma tail is often blue because it contains carbon monoxide ions. A few plasma tails can extend 100 million miles (160 million km) into space.

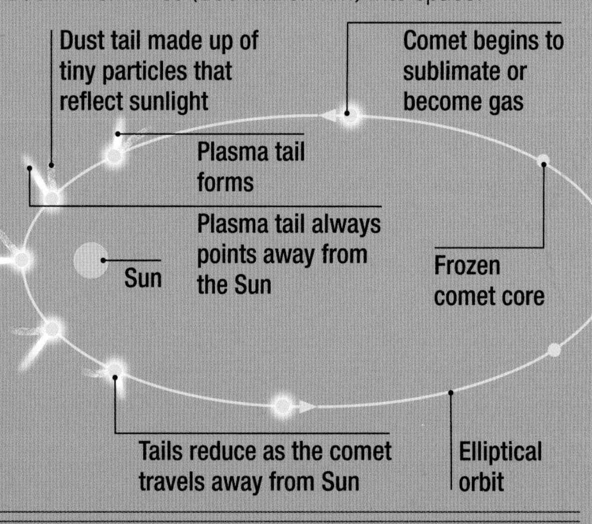

Many **comets** are **formed** in the **Oort Cloud** and **Scattered Disk**, two of the **outermost regions** of our **solar system**.

every 6.88 years — **Holmes**
Last here in 2014, expected next 2021

every 75.3 years — **Halley's**
Last here in 1986, expected next 2061

every 133.3 years — **Swift-Tuttle**
Last here in 1992, expected next 2126

every 2,380 years — **Hale–Bopp**
Last here 1997 expected next 4380

every 8,000 years — **Lovejoy**
Last here 2015, expected next 10,015

Regular visitors

In 1705, the astronomer Edmund Halley concluded that reports of a comet approaching Earth in 1531, 1607, and 1682 were in fact orbits of the same comet. He correctly predicted that the comet would return in 1758, and Halley's comet was named for him. Comets that have orbital periods of less than 200 years are called "short-period comets," and those with orbital periods of more than 200 years are named "long-period comets."

Doom-monger

Around 1509, ten years before the arrival of the all-conquering Spanish Conquistadors, it is said that Montezuma, ruler of the Aztecs, saw a "fire plume" in the sky. This was only the first of eight omens that he thought predicted the end of the empire and his own death. In August 1519, Hernán Cortés marched on the capital city of Tenochtitlan and Montezuma's fears were realized.

SATURN

GAS GIANT

Saturn is a giant slushy orb of gas and liquids. Clouds of ammonia ice sit above layers of frozen water, underneath which are cold hydrogen and sulfur ice mixtures. Below these, the planet's powerful gravity crushes the hydrogen so hard that it turns to metal. All this before we reach Saturn's blazing hot core.

Average distance from Sun: 886 million miles (1.43 billion km) / 9.5 AU

First record: Assyrians, 8th century BCE; first seen through a telescope by Galileo, 1610

Visited by: Pioneer 11 (1979), Voyager 1 (1980), Voyager 2 (1981), Cassini (2004)

Speed of orbit round the Sun: 21,562 mph (34,701 km/h)

Equatorial circumference: 235,298 miles (378,675 km)

Gravity: 106% of that on Earth

Year: 10,756 Earth days (more than 29 Earth years)

Temperature: –288°F (–178°C)

Atmosphere: mainly hydrogen, helium, plus traces of methane, ammonia, ethane

Known moons: 62 (of which 9 are still to be named)

Rings: 30+ in 7 groups

Beauty revealed

Second only in size to Jupiter, 764 Earths would fit inside Saturn. Seen through a telescope, the planet appears to be a pale yellow. For more than a decade, Cassini studied Saturn and its rings. It is only from the detailed pictures that the probe had relayed back to Earth that we have been able to see the planet in all its glory.

Saturn turns on its **axis** once every **10 hours** and **39 minutes**.

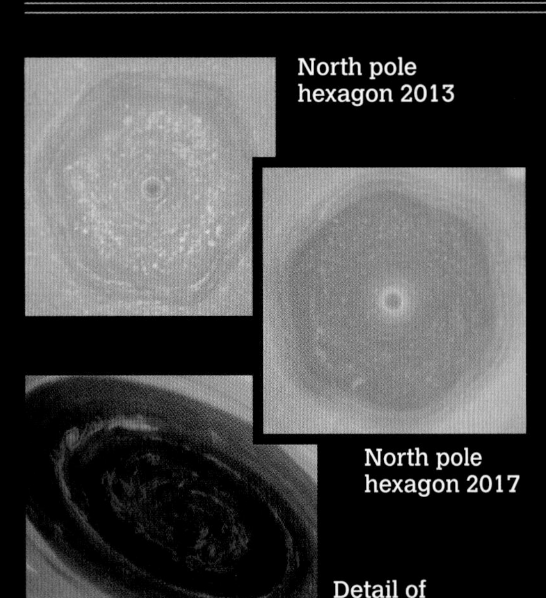

North pole hexagon 2013

North pole hexagon 2017

Detail of storm's center

The north pole

The odd, hexagonal cloud pattern (above right and left) at the north pole was first discovered in 1988 by scientists examining Voyager images. At its center is a hurricane about 20,000 miles (32,000 km) across that extends about 60 miles (100 km) down into the atmosphere. Nothing like the hexagon has been seen anywhere else so far.

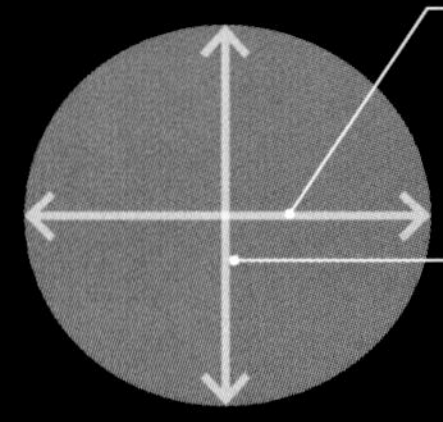

Equatorial radius
37,449 miles (60,268 km)

Polar radius
33,780 miles (54,364 km)

In a flat spin

Saturn is not round. It spins so quickly on its axis that the planet flattens itself, causing the equator to bulge outward.

Cloud pattern

The different makeup of the layers of clouds on Saturn cause a velocity difference between the layers. This is known as the Kelvin-Helmholtz instability and causes a pattern of clouds that are sometimes seen on Earth.

Cloud pattern on Earth

Colorful planet

Streaking out from Saturn's south polar clouds are curtains of aurorae. In this composite of near-infrared images from Cassini, the aurorae are bright green. The sunlight reflected off Saturn's rings appears blue and heat from the planet's interior is a deep red. The dark spots and bands are clouds.

Huge storm on Saturn

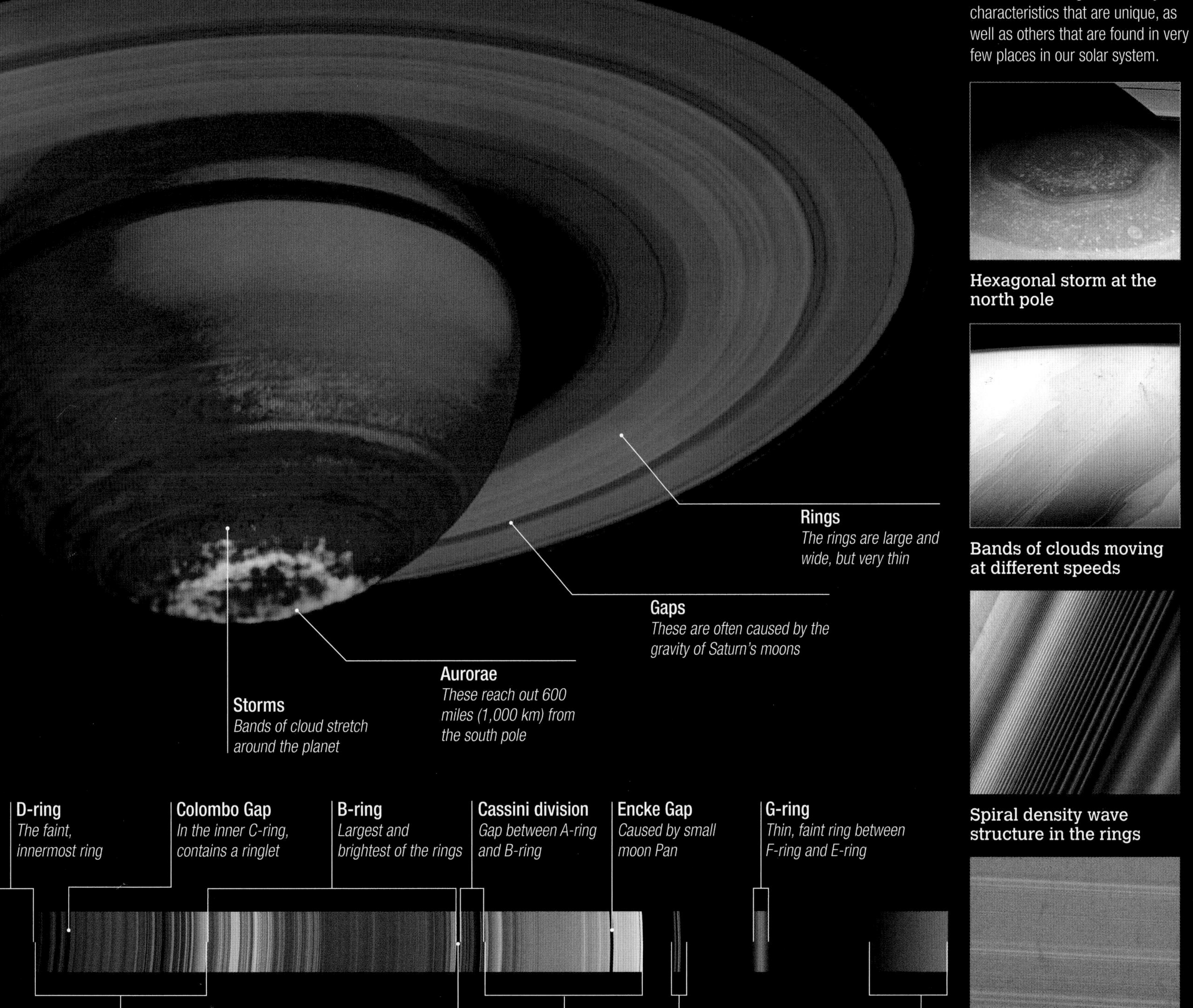

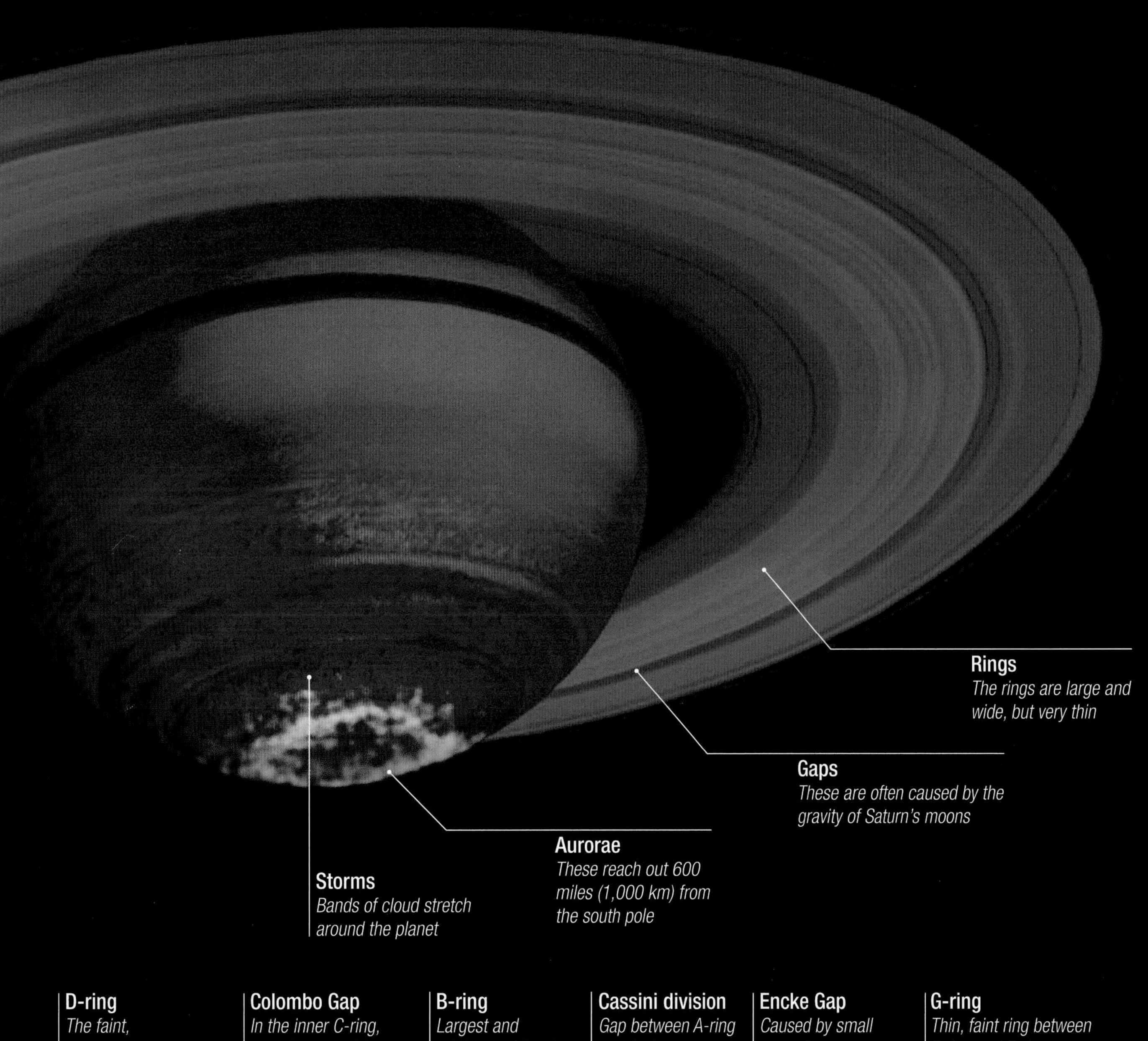

Features of Saturn

Saturn and its rings have many characteristics that are unique, as well as others that are found in very few places in our solar system.

Hexagonal storm at the north pole

Bands of clouds moving at different speeds

Spiral density wave structure in the rings

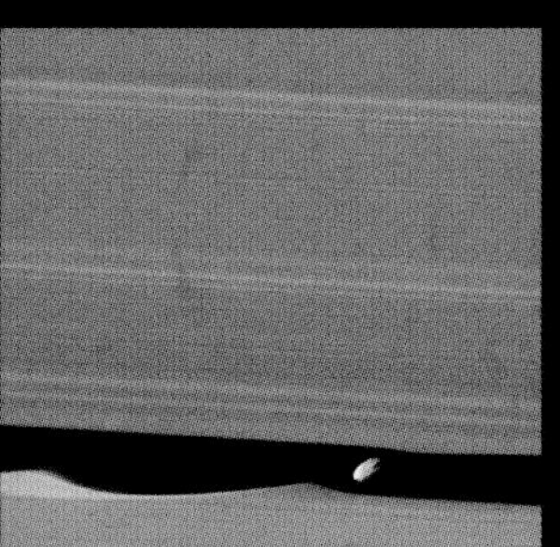

Wavemaker moon Daphnis forming gap in the rings

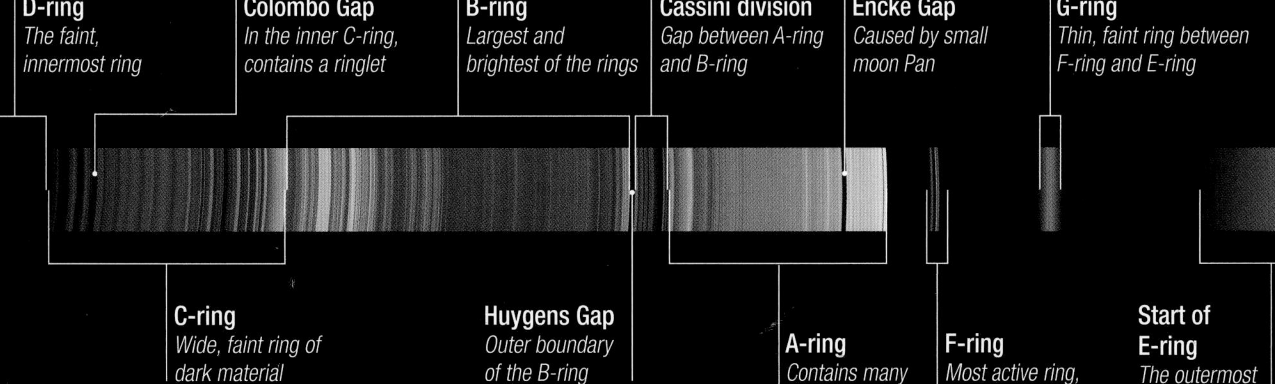

The rings

Saturn has the largest ring system of any planet, and it is one of the most beautiful sights to be seen. There are seven groups of rings made of ice and rock. Astronomers have named them from A to G and there are some gaps between that are named more imaginatively.

Looking back

On July 19, 2013, Cassini moved into Saturn's shadow and imaged the planet backlit by the Sun. The image shows seven of Saturn's moons, its inner rings, and also Earth, the tiniest of bright blue dots seen here between the outer, thin G ring and the faint smoky E ring. Earth is on the right at about 4 o'clock from the planet itself. This is only the third time that our home planet has been imaged from the outer solar system. Cassini took 323 wide- and narrow-angle pictures in just over four hours. This final mosaic uses 141 of them.

SATURN'S MOONS

NATURAL SATELLITES

Saturn has 62 moons, some of which were found by the Cassini probe during its many years in the planet's orbit. The largest moons are huge globes of ice and rock, including Titan with its freezing nitrogen atmosphere that creates an orange mist over a surface pitted by channels of liquid methane.

Quintet of moons
Five of Saturn's moons are in this image—Janus on the far left, Pandora in the rings with Enceladus above at center, then Mimas near the much larger Rhea.

Sinlap
This relatively "fresh" crater is 50 miles (80 km) across.

Through the haze
The use of infrared mapping allowed Cassini to penetrate the clouds that surround Titan and show its surface for the first time. What can be seen here are dune-filled terrains named Fensal (to the north) and Aztlan (to the south).

Distance from the Sun: 886 million miles (1,427 million km)

Orbital distance from Saturn: 759,000 miles (1.2 million km)

Orbit time: 16 days

Diameter: 3,200 miles (5,150 km)

Age of surface: 50 million years

Surface temperature: –180°F (–292°C)

Discovered by: Christiaan Huygens on March 25, 1655

Atmosphere: nitrogen, methane; traces of ammonia, argon, ethane

TITAN

LARGEST MOON

Saturn's largest moon seems similar to early Earth, and even has seasons, each lasting 7.5 years. It is the only place apart from our planet that has liquids flowing across its surface. Methane and ethane rain falls from its clouds.

Cloud covered
Titan is the only moon in our solar system that has a thick atmosphere—its atmospheric pressure is about the same as is found at the bottom of a swimming pool on Earth. Its outer atmosphere is full of hydrocarbons, and scientists think that it may be a laboratory for the type of organic chemistry that preceded life on Earth.

ENCELADUS

SNOWBALL SATELLITE

A vast, hidden, liquid reservoir fuels 101 geysers that spurt out jets of water ice and vapor at 800 mph (1,290 km/h), creating a ring of fine ice dust around this moon that contributes to Saturn's E-ring.

Distance from the Sun: 886 million miles (1,427 million km)

Orbital distance from Saturn: 148,000 miles (238,000 km)

Orbit time: 1.37 days

Diameter: 310 miles (500 km)

Age of surface: 100 million years

Surface temperature: –330°F (–201°C)

Discovered by: William Herschel on August 28, 1789

Atmosphere: thought to have a significant atmosphere

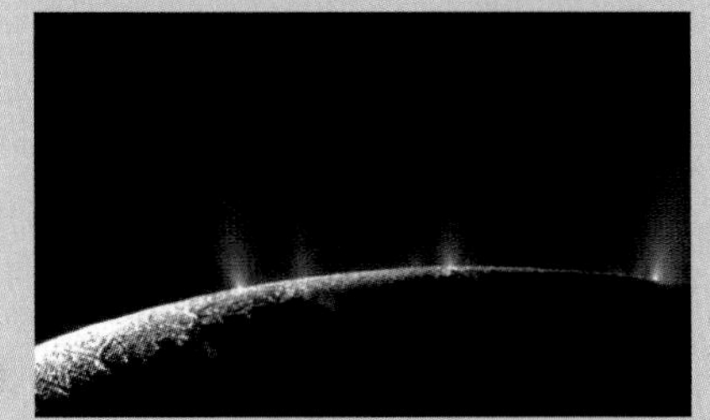

Active geysers
Cassini first spotted a series of geysers blasting water ice and vapor from the surface of the moon in 2005. This image, taken in 2010, shows several geysers erupting from fractures that cross the moon's south polar terrain. In total 101 geysers were discovered.

Ice-covered ocean
The observations of the geysers led to research into the moon's makeup. Cassini scientists mapped the position of features across hundreds of images. They noticed that the moon wobbles as it orbits Saturn and have concluded that this is because its outer icy shell covers a global ocean, which in turn feeds the geysers.

Ice planet
Tethys is heavily scarred, cold, and airless. It is made up almost entirely of water ice and covered in impact craters, one of which, Odysseus, has a diameter that is nearly two-fifths the size of Tethys itself.

Shining bright
The icy surface is highly reflective, so the floor of the many craters shine brightly.

TETHYS

BRIGHT MOON

The much-weathered surface of Tethys has a giant crater 250 miles (400 km) across that must have been blasted out when the moon was still forming—anything later would have shattered it. A deep valley 1,200 miles (2,000 km) long may have been caused by the same collision, or the expansion of internal liquids as they froze.

Distance from the Sun: 886 million miles (1,427 million km)

Orbital distance from Saturn: 183,000 miles (295,000 km)

Orbit time: 1.89 days

Diameter: 662 miles (1,066 km)

Age of surface: 20–180 million years

Surface temperature: –305°F (–187°C)

Discovered by: Giovanni Cassini on March 21, 1684

Atmosphere: none

DIONE

ICY BODY

Usually, the forward-facing side of a moon suffers most crater damage. But Dione is a puzzle, because it is much more weathered on the opposite hemisphere. Scientists believe it suffered a major collision that spun the whole moon around.

Pock-marked
The surface of this icy moon is covered with impact craters. The large, multi-ringed basin Evander, 220 miles (350 km) wide, is bottom right of this image.

Distance from the Sun: 886 million miles (1,427 million km)

Orbital distance from Saturn: 234,000 miles (377,400 km)

Orbit time: 2.7 days

Diameter: 698 miles (1,123 km)

Age of surface: 50 million years

Surface temperature: –302°F (–186°C)

Discovered by: Giovanni Cassini on March 21, 1684

Atmosphere: a thin layer of oxygen ions has been detected

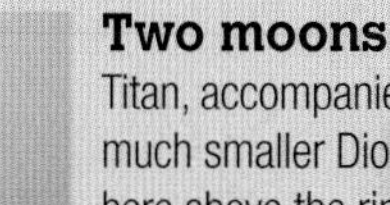

Two moons
Titan, accompanied by the very much smaller Dione are shown here above the rings of Saturn. This view looks toward the northern, sunlit side of the rings. Cassini's narrow-angle camera was at this point approximately 1.4 million miles (2.3 million km) from Titan and 2 million miles (3.2 million km) from Dione.

Tectonic forces may have **forged** towering **ice cliffs** on **Dione**.

OTHER MOONS

MOBILE DEBRIS

Some of Saturn's moons are small and oddly shaped, possibly comets captured by the planet's powerful gravity. Others were probably formed when a giant moon broke up and left the debris that forms Saturn's rings.

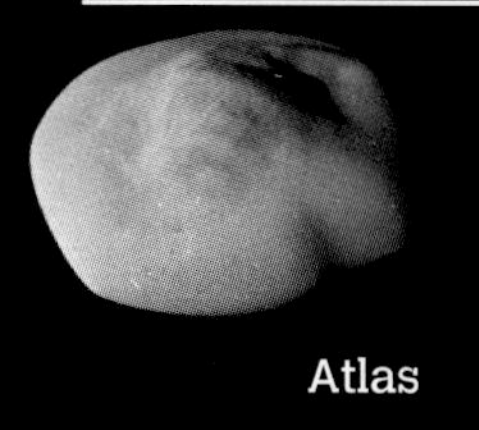

Atlas

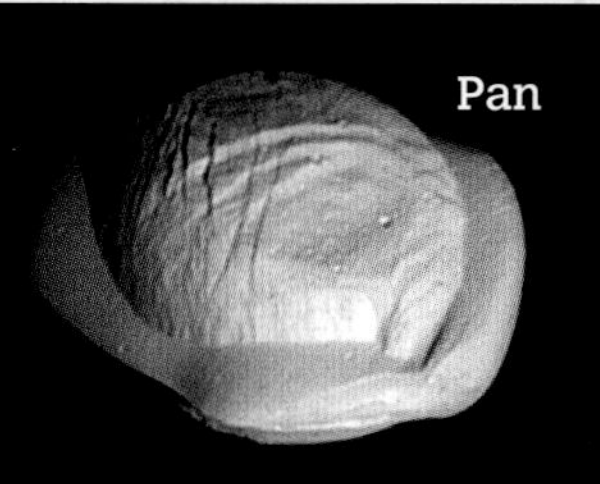

Pan

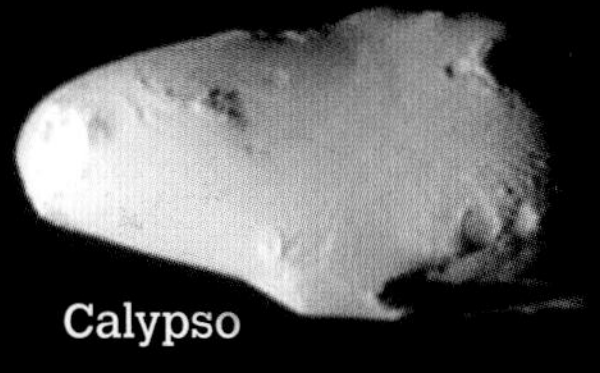

Calypso

Phoebe

Helene

Janus

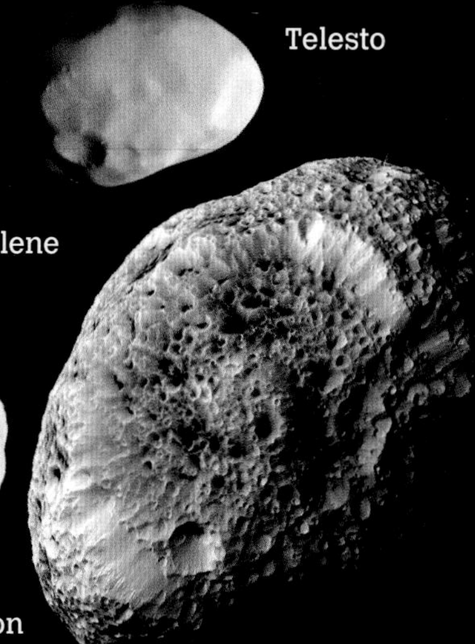

Telesto

Hyperion

DEATH OF A PROBE

CASSINI'S ODYSSEY

The Cassini spacecraft made a seven-year journey that took in Venus—twice—Earth and Jupiter, using their gravity to slingshot itself across the solar system. Then it spent 13 years touring Saturn and its moons before self-destructing in the planet's atmosphere to avoid contaminating these outer worlds.

Huygens

Cassini

Namesakes

NASA's Cassini probe was named for Italian astronomer and engineer Giovanni Cassini (1625–1712). ESA's Huygens lander remembers the Dutch scientist and mathematician Christiaan Huygens (1629–1695).

Cassini being assembled in Pasadena, California

The start of it all

The Cassini–Huygens project, usually called Cassini, was a collaboration between NASA, ESA, and the Italian space agency Agenzia Spaziale Italiana (ASI). The mission was to send the probe Cassini, and the lander Huygens, to study Saturn and its moons. They were launched aboard a Titan IVB/Centaur rocket.

Cassini lifts off from Cape Canaveral

Cassini was **large** for a **probe**, about the same **size** as a **30-passenger** school **bus**.

Huygens lander

Cassini had 12 high-tech instruments capable of 27 scientific investigations. When Huygens landed on Saturn's moon, Titan, it was the first probe to land on a world in the outer solar system.

RPWS system

This instrument detected radio and plasma waves with a suite of antennas and sensors.

445 Newton engine

The main engine burned fuel for maneuvers. An identical second engine acted as backup.

Magnetometer boom

The most sensitive measuring instruments were mounted on long booms, away from the craft.

Huygens probe

The probe was attached to, and launched from, Cassini.

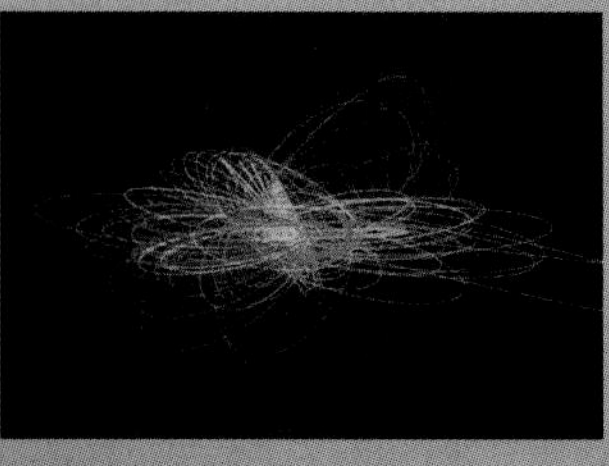

Round and round

The incredible journey that Cassini has been on is shown in this graph of its orbits around Saturn and its moons. Saturn is in the center.

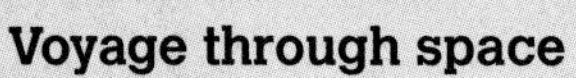

Voyage through space

Cassini used the gravities of Earth, Venus, and Jupiter to pick up speed for the trek out to Saturn. It toured its moons, sent Huygens onto the biggest one, and took a multitude of snapshots of the outer edges of the solar system. What a spacecraft!

- **Oct. 15, 1997** — *Launch from Cape Canaveral*
- **Apr. 25, 1998** — *First Venus flyby*
- **Dec. 1999** — *Reaches Asteroid Belt*
- **Dec. 29, 2000** — *Approaches Jupiter*
- **Apr. 7, 2004** — *Observes storms merging on Saturn*
- **Dec. 23, 2004** — *Huygens probe detatches over the moon Titan*
- **Jan. 13, 2005** — *Huygens lands on Titan*

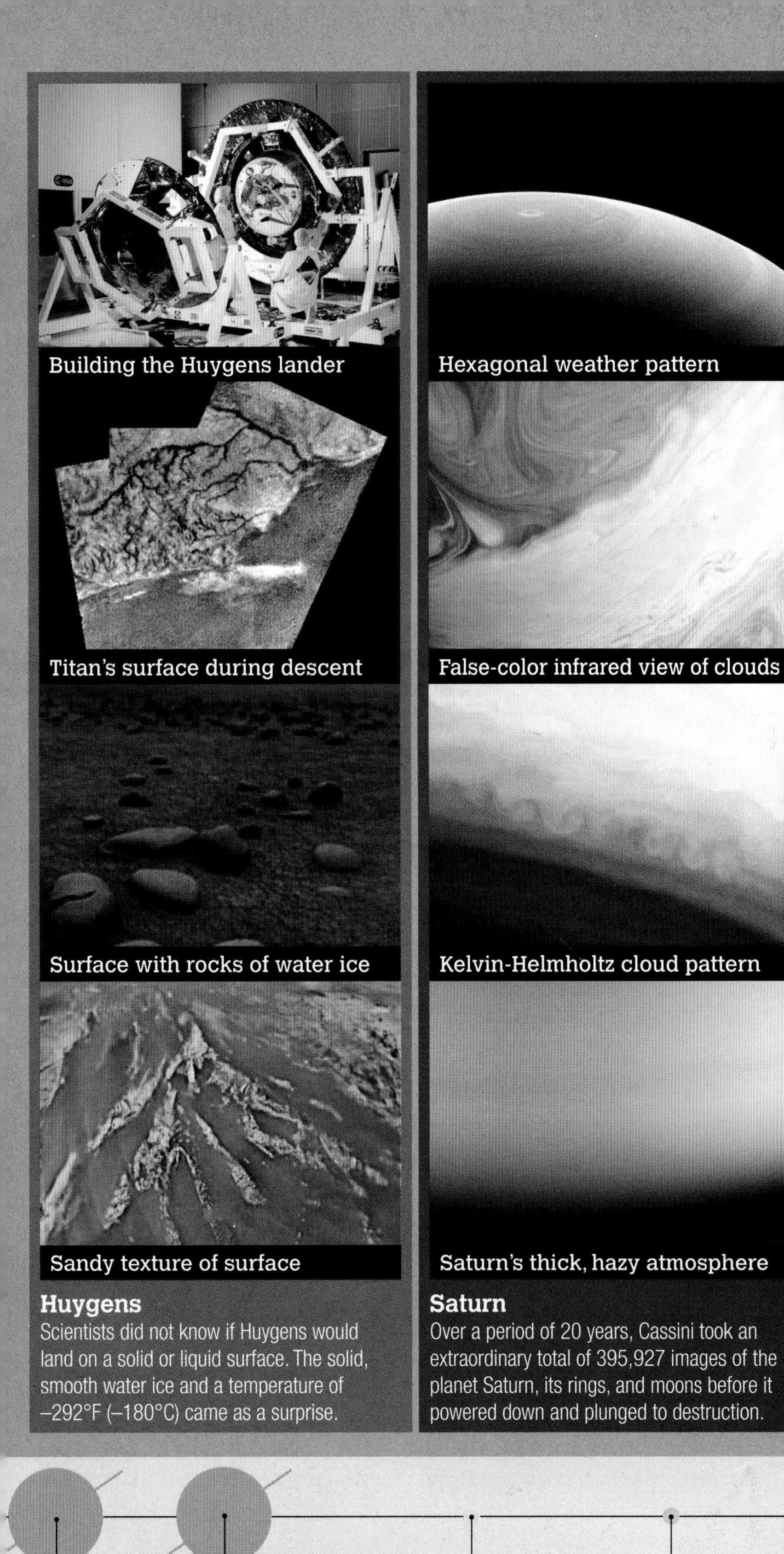

Building the Huygens lander

Titan's surface during descent

Surface with rocks of water ice

Sandy texture of surface

Huygens
Scientists did not know if Huygens would land on a solid or liquid surface. The solid, smooth water ice and a temperature of –292°F (–180°C) came as a surprise.

Hexagonal weather pattern

False-color infrared view of clouds

Kelvin-Helmholtz cloud pattern

Saturn's thick, hazy atmosphere

Saturn
Over a period of 20 years, Cassini took an extraordinary total of 395,927 images of the planet Saturn, its rings, and moons before it powered down and plunged to destruction.

Titan's upper atmosphere

Daphnis in the Keeler Gap

Cratered surface of Epimetheus

Pandora next to the F-ring

Moons
Cassini revealed Saturn's moons in glorious detail. Its observations of Enceladus in particular have opened scientists' minds to the possibility of life elsewhere in the solar system.

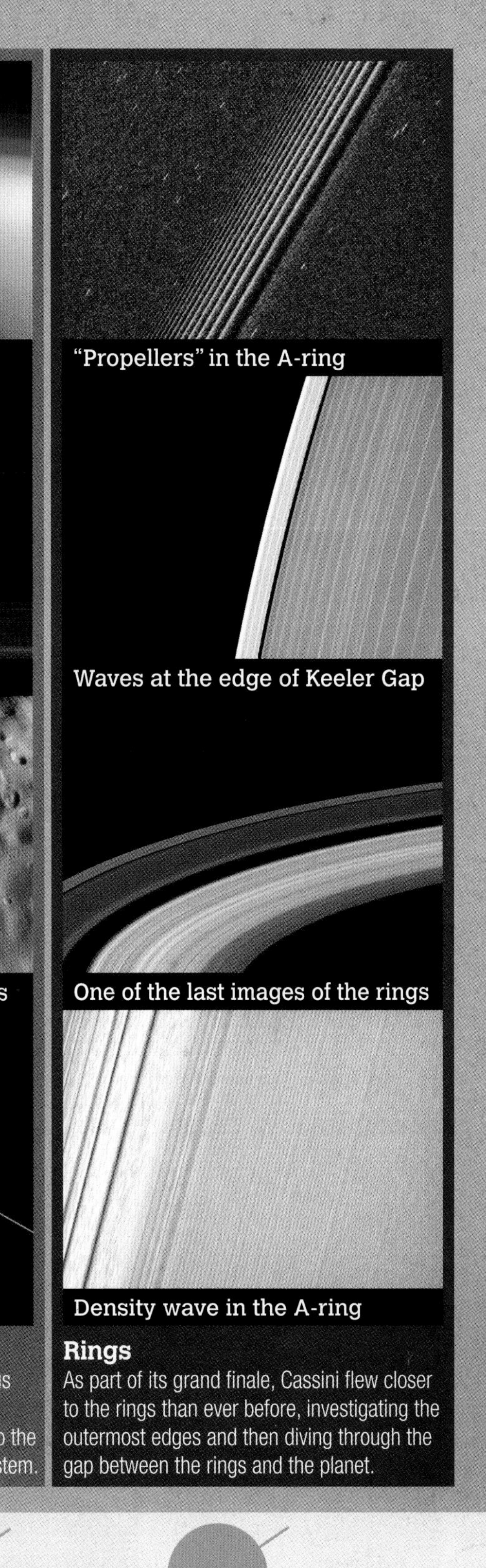

"Propellers" in the A-ring

Waves at the edge of Keeler Gap

One of the last images of the rings

Density wave in the A-ring

Rings
As part of its grand finale, Cassini flew closer to the rings than ever before, investigating the outermost edges and then diving through the gap between the rings and the planet.

Sept. 14, 2006
New rings around Saturn identified

Dec. 4, 2010
Monster storm tracked in Saturn's northern hemisphere

July 27, 2014
101 geysers identified erupting on moon Enceladus

May 31, 2015
Close approach to Hyperion reveals impact-scarred surface

Nov. 29, 2016
Begins inclined orbits to carry it closer to Saturn

Apr. 26, 2017
First of 11 dives through gap between Saturn and its rings

Sept. 15, 2017
Final plunge into Saturn's atmosphere

URANUS

GAS GIANT

An incredibly large object smashed into Uranus early in its history. The giant, icy, and windswept planet tumbled into a 98-degree tilt so that it now rolls almost on its side around the Sun. It takes 84 years to orbit, and the poles have 42 years of daylight followed by 42 years in the dark. Uranus' appearance does not give anything away—its top layer of methane hides its features from view.

Average distance from Sun: 1.78 billion miles (2.87 billion km) / 19.2 AU

First record: William Herschel, March 13, 1781

Visited by: Voyager 2 (1986)

Speed of orbit round the Sun: 15,209 mph (24,477 km/h)

Equatorial circumference: 99,787 miles (160,592 km)

Gravity: 90% of that on Earth

Year: 30,687 Earth days (84 Earth years)

Temperature: –357°F (–216°C)

Atmosphere: hydrogen, helium, methane

Known moons: 27

Rings: 11 inner rings and 2 outer rings

Elemental name

In 1789, German chemist Martin Klaproth was examining mineral samples from a silver mine when he discovered uranium. He named this fascinating new element for the planet Uranus, which had only been found eight years before.

Uranium ore

Big blue

Unseen without a telescope, Uranus appears a hazy blue because the methane gas in its upper atmosphere absorbs red light. The cool color echoes its coldness—temperatures in its cloud tops can be as low as –370°F (–224°C), making it the coldest atmosphere of any planet. Underneath seems to be a warm, slush-like mixture of water, ammonia, and methane thousands of miles deep.

Uranus has an **icy mantle** that **surrounds** its **rock** and **nickel–iron** alloy **core**.

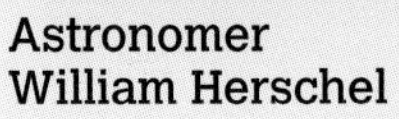

Astronomer William Herschel

Herschel's observatory

Is it a comet?

On March 13, 1781, British astronomer William Herschel was surveying all the stars that were of at least magnitude 8 (slightly too faint to see with the naked eye). He spotted a very faint object that he thought at first was a comet. Later, he realized it was Uranus, the first new planet discovered since ancient times.

Faint ring discovery

At the top of this image is the planet's—very faint—tenth ring, orbiting Uranus at a distance of about 30,000 miles (50,000 km). The ring was discovered by Voyager 2 in January 1986.

Storm

Uranus' ring system

Image taken by the Keck Telescope

Got a ring

In March 1977, scientists observing the star SAO 158687 noticed that at least five rings of Uranus were getting in the way. In 1986, when Voyager 2 gave Earth a closer view, it was established that there are 13 rings.

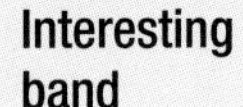

Interesting band

Stormy pole

Hidden storms

By combining multiple images of Uranus in near-infrared, the Keck telescopes were able to give us a glimpse of the extraordinary weather systems that lurk beneath the methane clouds. Great circulating bands of cloud race around the planet at up to 560 mph (900 km/h).

Color enhancement

False-color and contrast enhancement images taken by Voyager 2 have revealed details that are too subtle to pick up in other ways. This image confirmed that the atmosphere of this planet, composed mainly of hydrogen and helium, circulates in the same direction as the planet rotates.

Sun rays on north pole

Uranus

Sun

21-year north pole summer

North pole in shade

21-year south pole summer

Long, hot summer

Earth has four seasons, caused by the fact that Earth's axis is tilted 23.5 degrees, and its orbit of the Sun is nearly circular. Uranus also has four seasons and a nearly circular orbit, but its tilt is 98 degrees and it has very different seasons. For the two 21-year summer and winter seasons, the hemispheres on Uranus are pointed either toward the Sun or away from it.

URANUS' MOONS

NATURAL SATELLITES

Voyager 2's quick flyby in 1986 and later observations from the Hubble telescope gave us major insights into the 27 moons of Uranus and identified a tenth ring. But they also posed mysteries about their features that scientists still struggle to explain. Most of the moons are named for characters from Shakespeare's plays, and two, Cordelia and Ophelia, are known as "shepherd moons" because they hold one of the planet's orbiting rings in place.

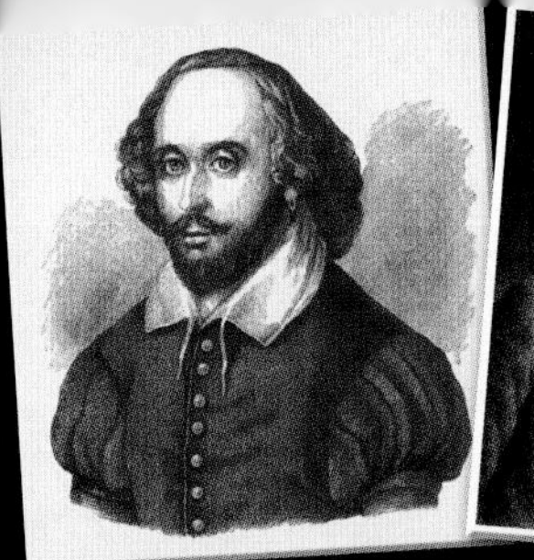

William Shakespeare

Alexander Pope

It's in the name

William Shakespeare was not the only inspiration for the moons' names. The name Umbriel was taken from the 18th-century poet Alexander Pope's writings.

MIRANDA

INNERMOST SATELLITE

Miranda has an oddly deform surface, with massive crum terraces, and giant fault lin This is probably caused by forces from Uranus squee and stretching the insides

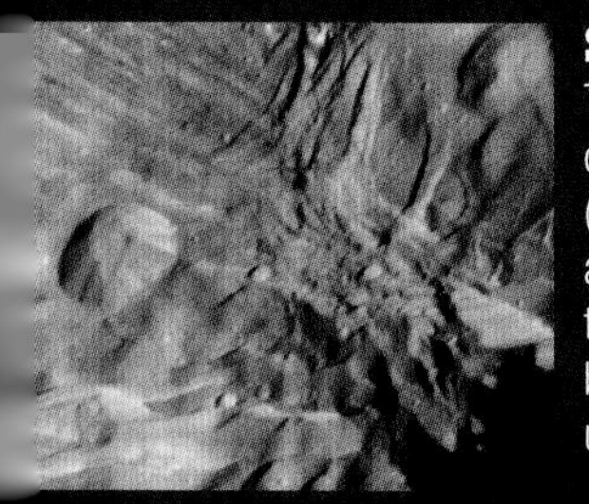

Surface features

This moon has giant fault canyons as deep as 12 miles (20 km)—12 times as deep as the Grand Canyon. Scientists think that the moon may have been shattered and reformed up to five times.

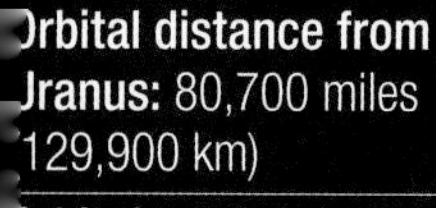

Orbital distance from Uranus: 80,700 miles (129,900 km)

Orbit time: 1.4 days

Diameter: 310 miles (500 km)

Surface temperature: –350°F (–213°C)

Discovered by: Gerard Kuiper on February 16, 1948

Atmosphere: none

Faithful daughter

The moon is named for the daughter of Prospero, exiled with her father in Shakespeare's *The Tempest*.

Cracked moon

Voyager 2 did the closest flyby of a moon wher it approached Miranda to get the boost it needec to reach Neptune. Its observations showed a moon unlike anything else yet seen in the solar system, with an extraordinary mixture of old and young surfaces.

ARIEL

BRIGHT SATELLITE

Studies of how light reflects from Ariel suggest that its surface is spongy because countless tiny meteorite strikes have turned the hard crust to soil. It is pitted with craters, with recent impacts wiping out the marks from larger, earlier collisions.

Orbital distance from Uranus: 118,640 miles (190,930 km)

Orbit time: 2.52 days

Diameter: 719 miles (1,158 km)

Surface temperature: –350°F (–213°C)

Discovered by: William Lassell on October 24, 1851

Atmosphere: none

Casting a shadow

The bright dot (left) is Ariel transiting Uranus, in an image taken by the Hubble Telescope in 2006. To the right of the moon is the dark spot that is Ariel's shadow.

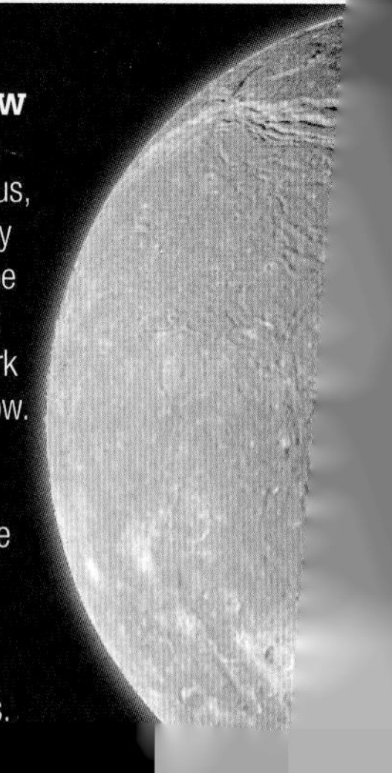

Marked moon

Ariel is mostly water ice and rock. Its surface is probably the youngest of all Uranus' moons. It is also thought to have had the most recent geological movements, its surface being criss-crossed by many valleys.

Uranus' family
This composite image from Voyager 2 shows the familiar blue planet Uranus with its five major moons. From smallest to largest, as they appear here, they are Ariel, Miranda, Titania, Oberon, and Umbriel. Voyager 2 also discovered 11 previously unseen moons.

UMBRIEL

DARKEST MOON

A mysterious moon—scientists believe Umbriel's dark appearance is caused by methane that has left a shadowy, carbon-rich residue.

Orbital distance from Uranus: 165,223 miles (265,970 km)

Orbit time: 4.14 days

Diameter: 750 miles (1,200 km)

Surface temperature: –325°F (–198°C)

Discovered by: William Lassell on October 24, 1851

Atmosphere: none

Strange ring
Umbriel is the darkest of Uranus' largest moons, reflecting only 16 percent of the light that strikes it. In 1986, Voyager 2 found an unexplained bright ring about 90 miles (140 km) across on its surface.

Orbital distance from Uranus: 362,880 miles (584,000 km)

Orbit time 13.5 days

Diameter: 720 miles (1,158 km)

Surface temperature: –325°F (–198°C)

Discovered by: William Herschel on January 11, 1787

Atmosphere: none

Enchantment in space
Oberon is the king of the fairies in Shakespeare's *A Midsummer Night's Dream*, and Titania (right) is his queen. Both moons were discovered by William Herschel and the names were suggested by his son John.

OBERON

OUTERMOST MOON

This is the most heavily cratered of Uranus' moons, suggesting it is the oldest. It is also much more red in color than its siblings.

TITANIA

LARGEST MOON

This is the largest moon and also the youngest because it is less cratered, although one large pit is 224 miles (360 km) across. Titania's crust is broken up by massive fault lines.

Orbital distance from Uranus: 271,104 miles (436,300 km)

Orbit time: 8.71 days

Diameter: 981 miles (1,578 km)

Surface temperature: –334°F (–203°C)

Discovered by: William Herschel on January 11, 1787

Atmosphere: none

Past activity
This moon has enormous valleys across its surface, some nearly 1,000 miles (1,600 km) long. These show that the moon has been geologically active.

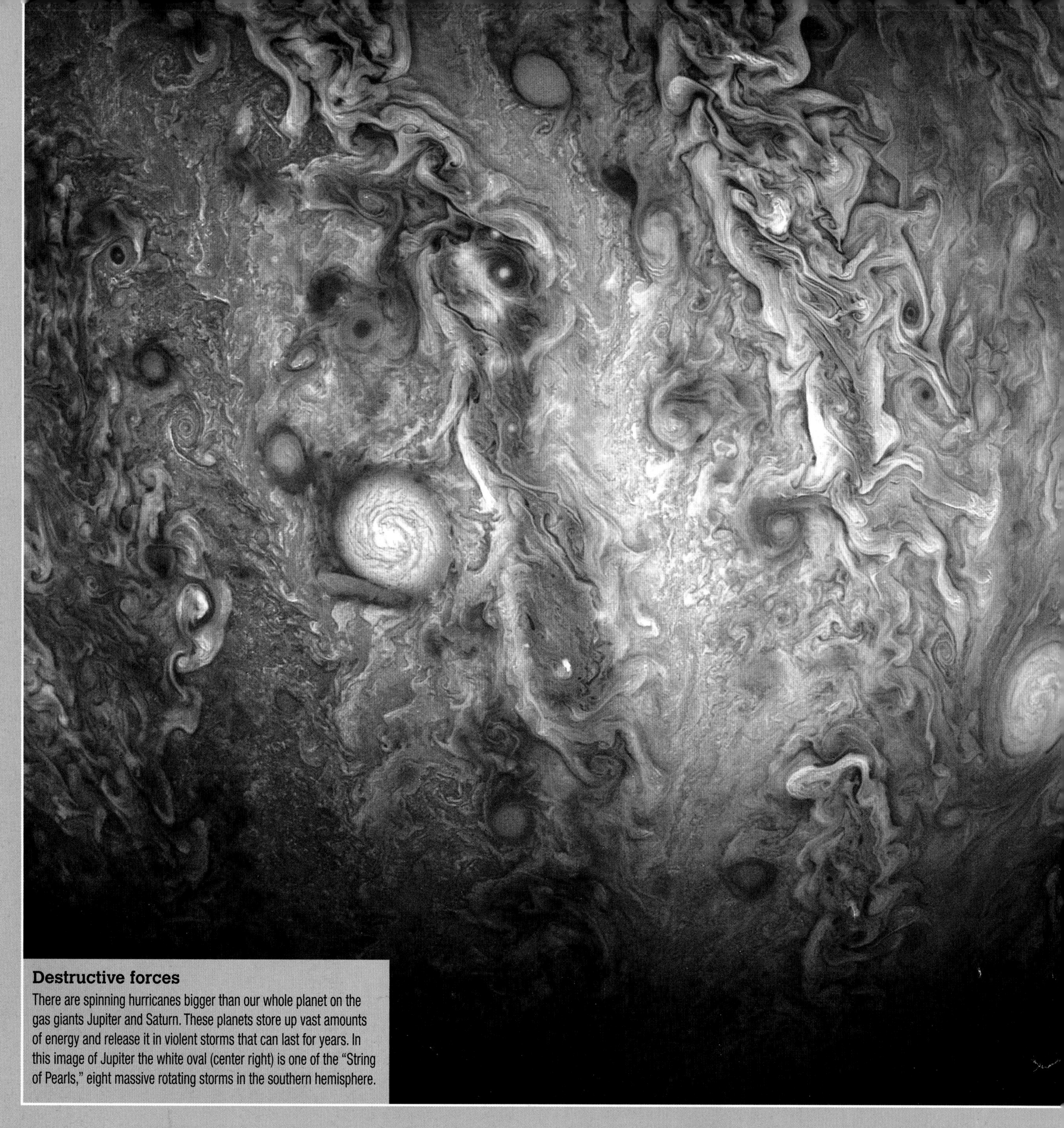

Destructive forces

There are spinning hurricanes bigger than our whole planet on the gas giants Jupiter and Saturn. These planets store up vast amounts of energy and release it in violent storms that can last for years. In this image of Jupiter the white oval (center right) is one of the "String of Pearls," eight massive rotating storms in the southern hemisphere.

PLANETARY WEATHER

On Earth, terrible hurricanes can wreck whole cities, but they look like a summer breeze compared to the weather on some other planets. Our storms are born over the warm seas and die out over the land. However, on planets with no continental crust, such as Jupiter, the violent tempests can carry on raging for years. Huge dust storms on Mars produce static electricity that blazes into huge forks of lightning.

The power of stormy weather

The central eye of a destructive Earth hurricane may be just 2 or 3 miles (3 or 5 km) across, but on Saturn it can measure about 2,500 miles (4,000 km) in diameter. And Neptune has the strongest winds of all, traveling faster than the speed of sound on Earth.

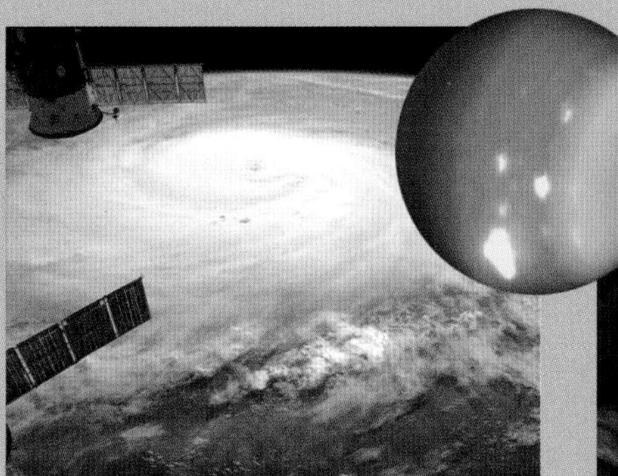

Hurricane Gonzalo seen from ISS over the eastern Atlantic Ocean

Storms on Uranus

Great Red Hole storm on Saturn

Huge counter-clockwise storm on Saturn

Dusty action

If a dust storm raged over the whole of North America for weeks, we would think it was the end of the world. On Mars, this happens several times a year—sometimes with the tempest growing big enough to blanket the whole planet in swirling red dust. However, the speed of the wind in these storms is only up to 60 mph (100 km/h), less than half the speed of some hurricane-force winds on Earth.

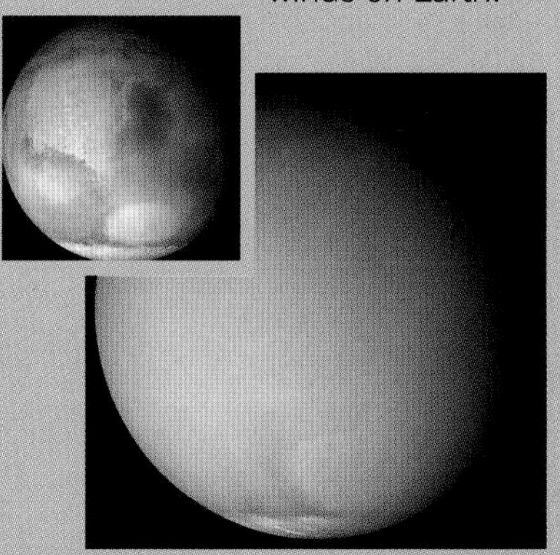

Whole planet dust storm on Mars

Dust storms on Mars—a major storyline in the movie *The Martian*

Dust devil photographed from Mars rover Curiosity

Precipitation

"Rain" and "snow" in space are not always water-based. There are "snowy" tops to Venus' mountains, capped with metallic galena and bismuthinite. On the moon Titan, it rains methane.

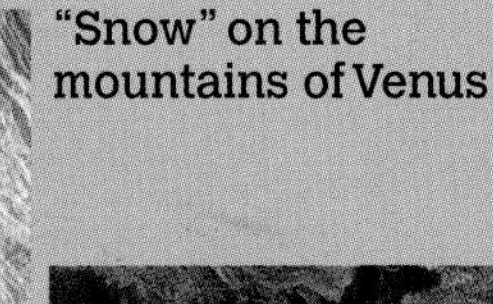

"Snow" on the mountains of Venus

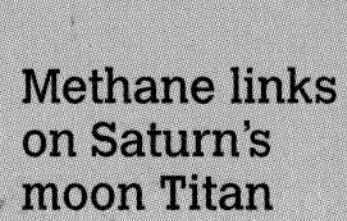

Methane links on Saturn's moon Titan

Neptune's **winds** travel at **1,118 mph (1,800 km/h)**.

Bright lights

Aurorae are caused by high-energy particles entering a planet's atmosphere. On Saturn and Jupiter, these brilliantly colored lights cover vast areas.

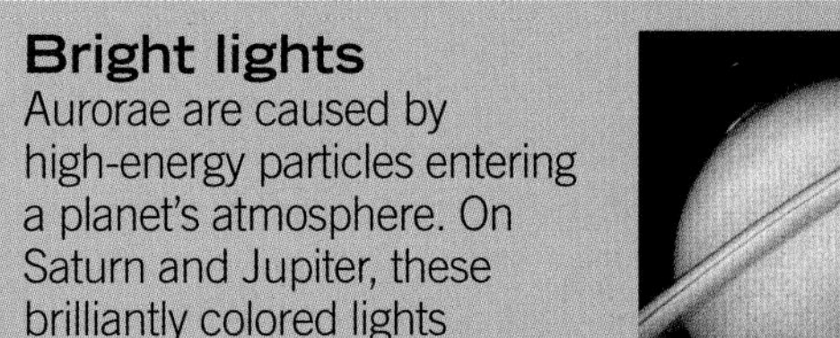

Aurora on Jupiter's southern hemisphere

Aurorae on Saturn's northern and southern hemispheres

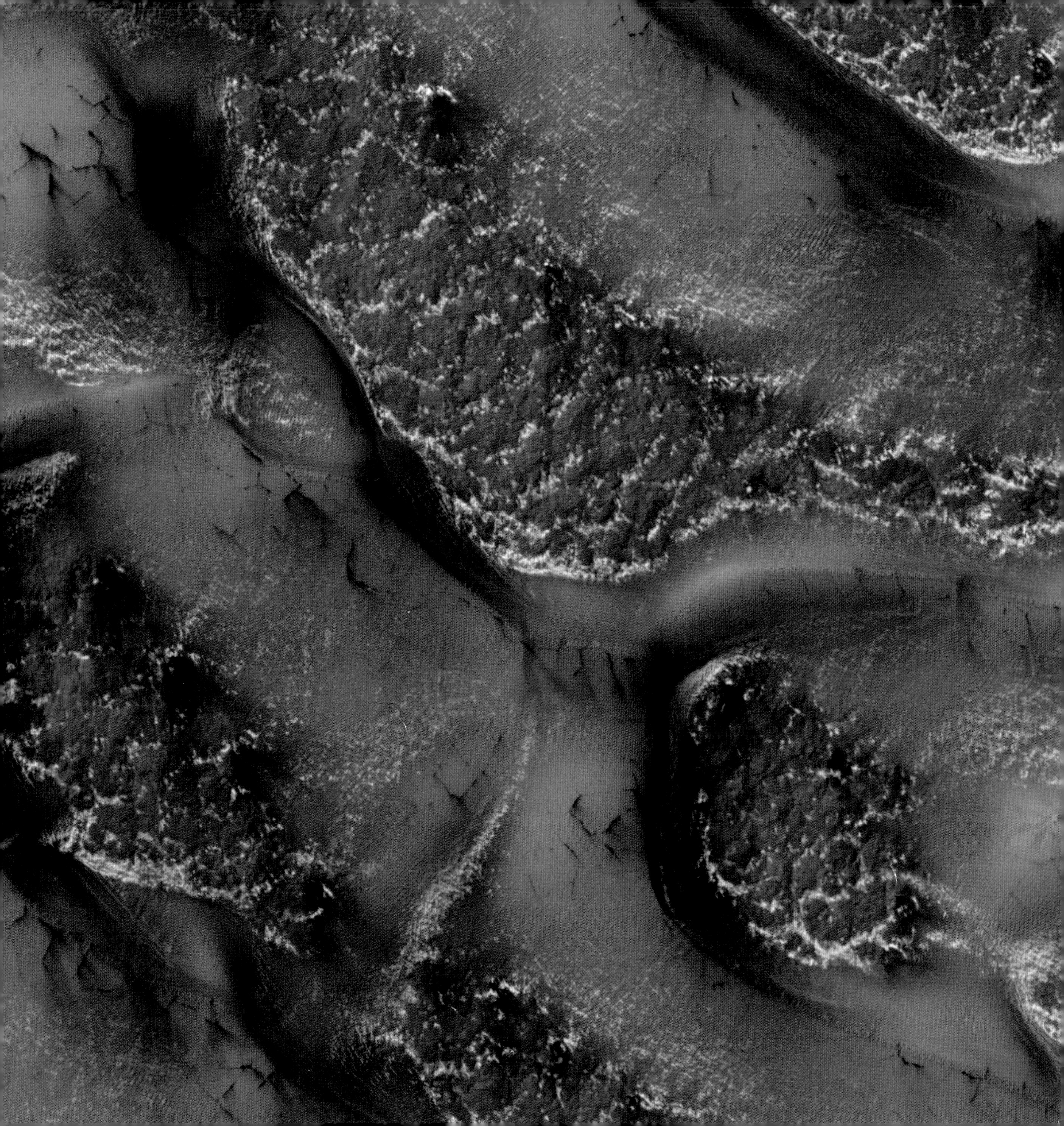

"Snow" on the red planet

In May 2017, the Mars Reconnaissance Orbiter took this picture of snow and ice covering dunes in the planet's northern hemisphere. Water-ice snowflakes coated in frozen carbon dioxide fall from clouds high in the planet's atmosphere and become trapped behind sheltered ridges. At night on the surface of Mars, temperatures can plunge as low as –195°F (–125°C). The drop in temperature combined with unstable air currents causes those carbon dioxide ice particles that are large enough to reach the surface as "snow".

NEPTUNE

GAS GIANT

Since its discovery in 1846, this most distant planet in our solar system has only completed one orbit of the Sun. Massive storms whirl around Neptune, pushed by 1,760 ft (536 m) per second winds that are fed by intense heat from the core. Despite this, Neptune has a chillingly extreme typical surface temperature of –392°F (–200°C).

Crescent planet
The crescents of Neptune and, below, its moon Triton were imaged by Voyager 2 at a distance of 3 million miles (4.86 million km) as the probe headed away from the planet.

Average distance from the Sun: 2.8 billion miles (4.5 billion km) / 30 AU

First record: Urbain Le Verrier, John Couch Adams, Johann Gottfried Galle on September 23, 1846

Visited by: Voyager 2 (1989)

Speed of orbit round the Sun: 12,158 mph (19,566 km/h)

Equatorial circumference: 96,683 miles (155,597 km)

Gravity: 114% of that on Earth

Year: 60,190 Earth days (165 Earth years)

Temperature: average –392°F (–200°C)

Atmosphere: hydrogen, helium, methane

Known moons: 14

Rings: a group of 5 very thin rings

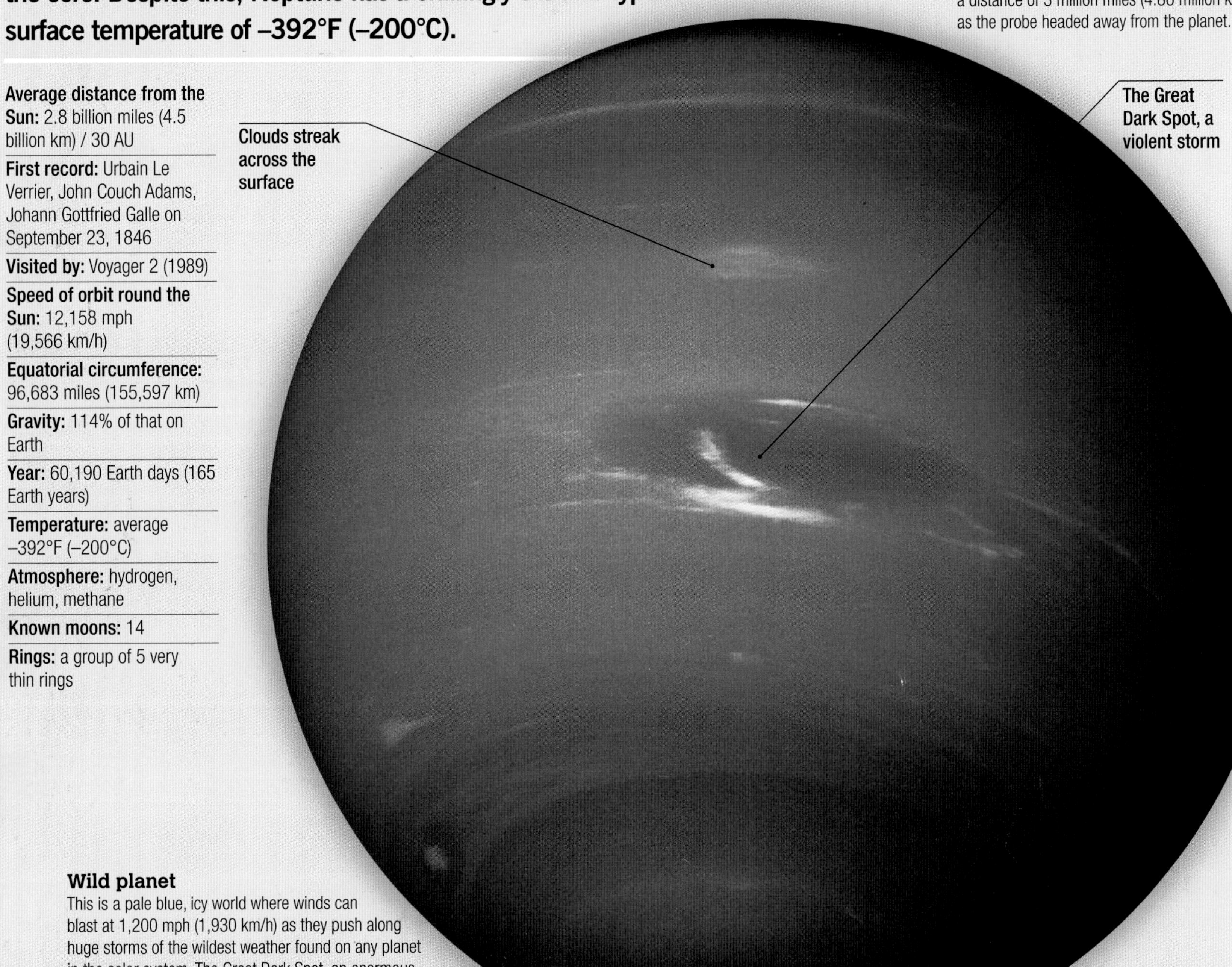

Wild planet
This is a pale blue, icy world where winds can blast at 1,200 mph (1,930 km/h) as they push along huge storms of the wildest weather found on any planet in the solar system. The Great Dark Spot, an enormous rotating storm system with winds of up to 1,370 mph (2,200 km/h) was first discovered in 1989 by Voyager 2.

Ring detail

Faint rings

At first it was thought that the planet's rings were incomplete, but Voyager 2 established that there are five of them. They are made up of 20–70 percent dust, ice particles, and rocks.

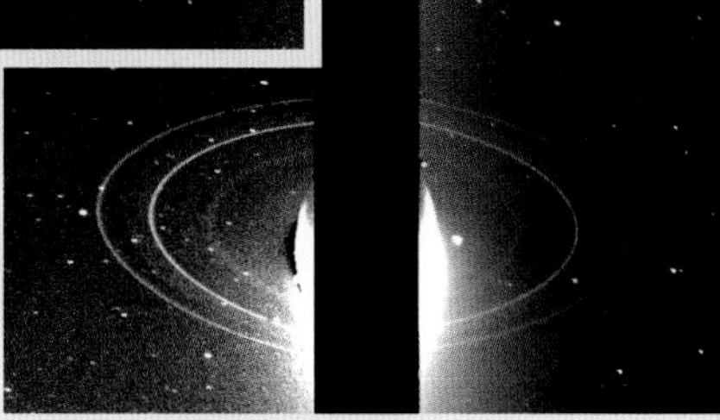

Ring system around Neptune

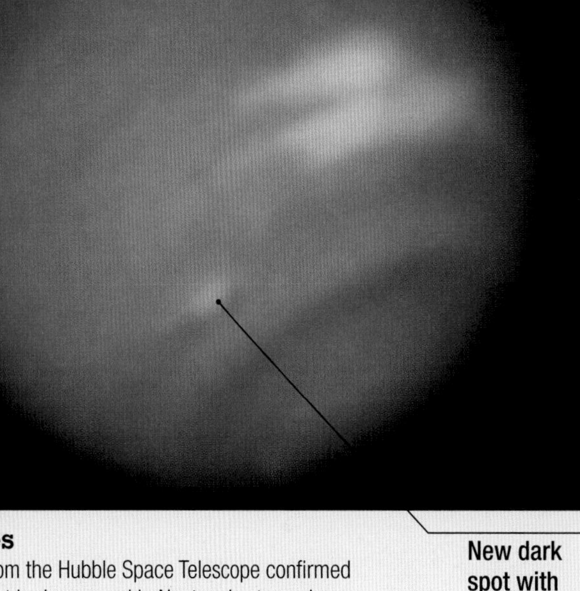

Climatic clues

In 2016, images from the Hubble Space Telescope confirmed that a new dark spot had appeared in Neptune's atmosphere. Storms on Neptune, unlike those on Jupiter, form and disappear again over a period of years, rather than centuries.

New dark spot with "companion" bright clouds

Features of Neptune

Fast-moving upper atmosphere clouds whirl around this blue planet, carrying storms raining methane ice with them.

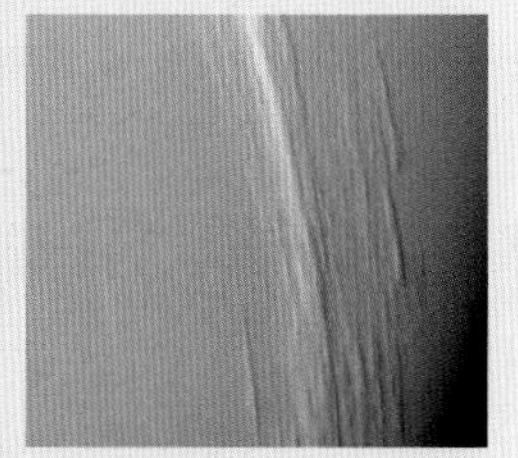

High clouds

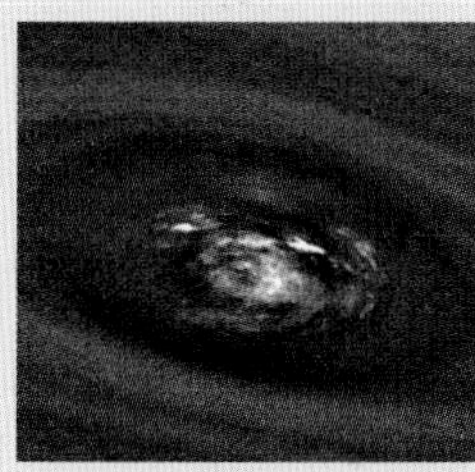

Huge storm systems

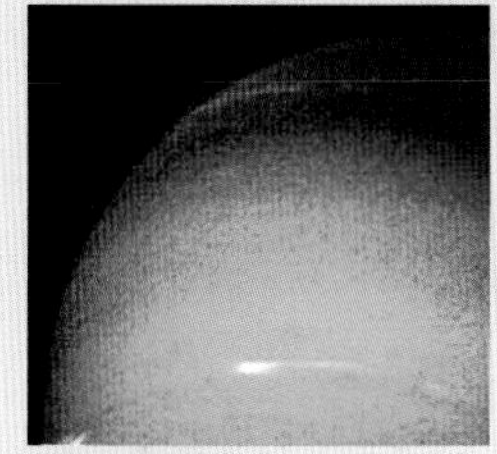

Methane in upper atmosphere

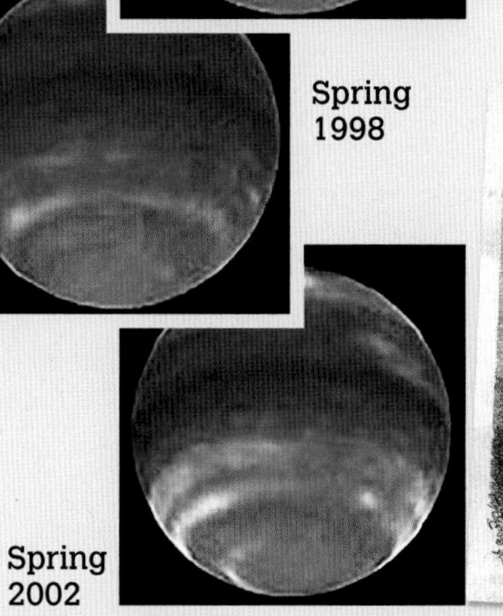

Spring 1996

Spring 1998

Spring 2002

Seasonal changes

There has been an increase in brightness over the years in the southern hemisphere. This indicates that the Sun is warming that part of the planet and proves there are seasons.

First sighting

In 1846, at the Berlin Observatory (above), German astronomer Johann Gottfried Galle used calculations made by Urbain Le Verrier, a French mathematician, to locate Neptune.

Single visitor

Voyager 2 arrived at Neptune in the summer of 1989, and gave the first sight of its moon Triton. The probe made a single flyby on its tour of the outer planets, taking pictures of the gas giant at a distance of just 3,000 miles (4,950 km).

Neptune is the **only** planet so far to have been **discovered** as a result of a **mathematical prediction**.

Coldest of temperatures

On Earth, the lowest natural temperature ever recorded is –128.6°F (–89.2°C) at the Vostok Station in Antarctica (above). On this ice giant, the temperature can reach –392°F (–200°C).

Continuing a tradition

The Romans named the five planets nearest the Sun for their most important gods. Urbain Le Verrier, who first identified this distant planet, chose Neptune, the Roman god of the sea.

NEPTUNE'S MOONS

NATURAL SATELLITES

Neptune's 14 dark and distant moons are pretty hard to spot. The first was found in 1846 and the latest discovered in 2013. The newcomer is known as S/2004 N1, a less romantic name than the sea gods and nymphs of the others. Triton is the only sphere, and five are probably comets sucked in by the gas giant's gravity.

Bright moons
This image of Neptune by Hubble shows bands of methane in its atmosphere and four of its brighter moons, Proteus (top), Larissa, Galatea, and Despina.

DESPINA

TINY SATELLITE

Only 90 miles (150 km) across, little Despina is slowly breaking up. It will one day end its eight-hour orbits and crumble into Neptune's rings.

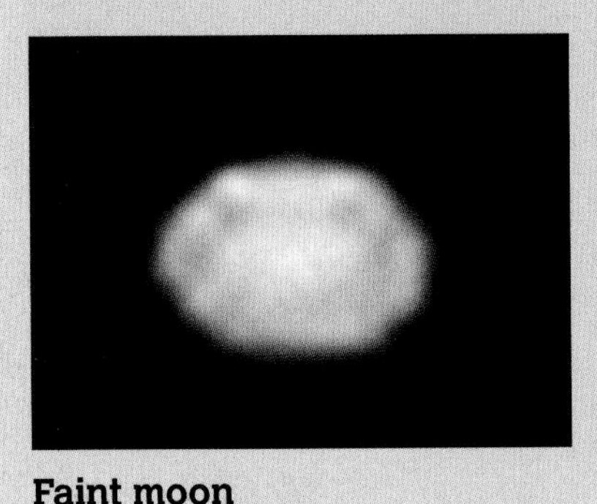

Faint moon
This irregularly shaped moon is inside Neptune's faint ring system. In mythology, Despina is one of the daughters of Poseidon, the Greek equivalent of the Roman god Neptune.

Orbital distance from Neptune: 32,600 miles (52,500 km)

Orbit time: 0.34 days

Diameter: 90 miles (150 km)

Surface temperature: –368°F (–222°C)

Discovered by: Voyager 2 July 28, 1989

Atmosphere: none

LARISSA

CROOKED MOON

Small, misshapen, and covered in craters, Larissa is part of Neptune's faint ring system. It will eventually either enter Neptune's atmosphere or break up into a planetary ring.

Orbital distance from Neptune: 45,700 miles (73,500 km)

Orbit time: 0.55 days

Diameter: 120 miles (190 km)

Age of surface: not known

Surface temperature: not known

Discovered by: William Hubbard, Larry Lebofsky, Harold Reitsema, and David Tholen on May 24, 1981, but officially found by the Voyager 2 science team in 1989

Atmosphere: none

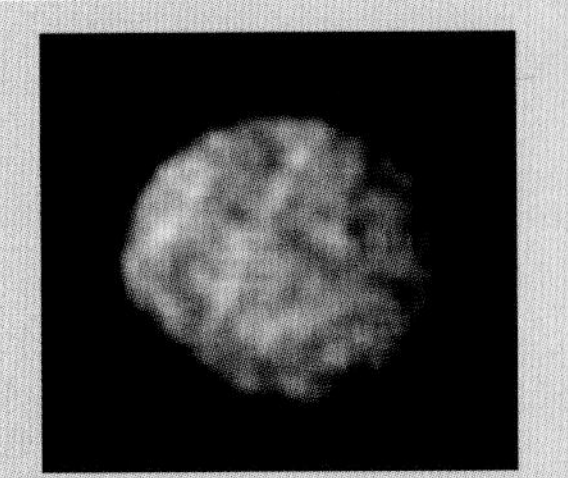

Best of the rest
Although difficult to see in its position near the ring system, it is thought that this moon is cold, icy, and heavily cratered. Some of the craters are 30 miles (50 km) across.

PROTEUS

DARK MOON

Proteus is dark and heavily cratered. Its surface only reflects 6 percent of the sunlight reaching it.

Orbital distance from Neptune: 73,100 miles (117,600 km)

Orbit time: 1.12 days

Diameter: 250 miles (400 km)

Age of surface: not known

Surface temperature: not known

Discovered by: Voyager 2 on June 16, 1989

Atmosphere: none

Shape changer
This moon, discovered in 1989, has been described as being "as dark as soot." One of the largest of Neptune's moons, scientists say that Proteus is probably as large as a satellite can get without being pulled into a sphere by its own gravity. It is named for the shape-changing Greek sea god.

PSAMATHE

OUTER SATELLITE

This moon, distant from Neptune, may have formed with another moon, Neso, when a larger body broke up.

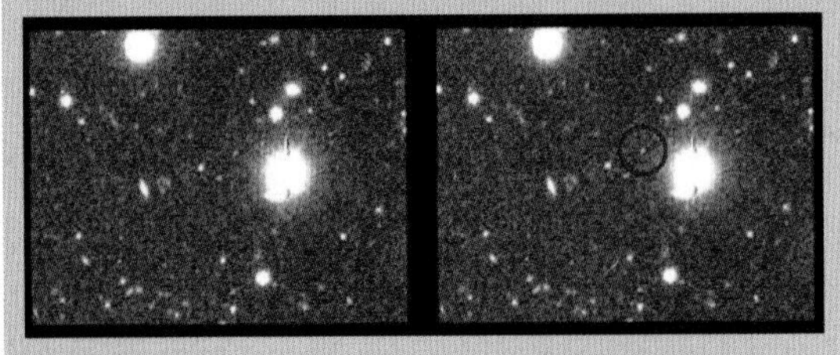

Distant view
This moon is small and very far away from Neptune—a tiny dot circled in red in this image from the Mauna Kea telescope.

Orbital distance from Neptune: between 16 and 42 million miles (25.7 and 67.7 million km)

Orbit time: 25 years

Diameter: 24 miles (38 km)

Age of surface: not known

Surface temperature: not known

Discovered by: Scott Sheppard, David Jewitt, and Jan Kleyna on August 29, 2003

Atmosphere: none

TRITON

LARGEST MOON

This frozen, rocky world has a thin atmosphere of nitrogen and methane, created when these gases were blasted upward from its many geysers, which are still active. With surface temperatures of –391°F (–235°C), it is a very cold moon.

Orbital distance from Neptune: 220,437 miles (354,759 km)

Orbit time: 5.9 days (retrograde)

Diameter: 1,680 miles (2,700 km)

Age of surface: about 50 million years

Surface temperature: –391°F (–235°C)

Discovered by: William Lassell on October 10, 1846

Atmosphere: mainly nitrogen with small amounts of methane

Ice moon

Triton is the largest of Neptune's moons and only slightly smaller than Earth's Moon. Like the Moon, it has a crater-pitted surface. But Triton is geologically active with geysers and cryovolcanoes, or ice volcanoes—the first to be discovered in our solar system. They spew out cold gases and ice, so the surface around them is constantly changing shape.

Smooth, volcanic plain

Triton is **one of** the **coldest places** in the **solar system**.

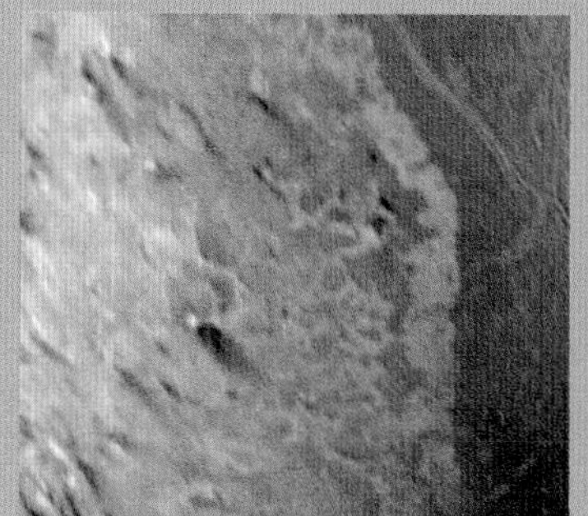

Volcanic vents

This image of Triton is a composite of pictures taken in 1989, when Voyager 2 was about 118,000 miles (190,000 km) away from the moon. The smooth, dark materials alongside the long, narrow canyons may be old volcanic deposits that were blown out of ancient vents.

Orbit plane and direction of orbit of Triton

Neptune

Orbit plane and direction of orbit of other moons

157° angle on inclination

Eccentric orbits

Triton is the only large moon in the solar system that circles in the opposite direction to its planet in a "retrograde orbit." Nereid, one of the outermost moons, also has an eccentric orbit—it is very stretched out.

Voyagers in space

The two spacecraft are identical—about the size of a van, weighing just 1,797 lbs (815 kg), and fitted with television cameras (now powered down), a radio receiver and transmitter, infrared and ultraviolet sensors (also powered down), magnetometers to measure the strength of magnetic fields, and plasma detectors, as well as sensors to detect cosmic rays and charged particles. They each carry a gold-plated disk with music, natural sounds such as whales and waves, 115 images of Earth, and greetings in 55 languages.

VOYAGERS THROUGH THE SOLAR SYSTEM

BY JEFFERSON HALL, MISSION FLIGHT DIRECTOR, VOYAGER TEAM

In 1972, NASA set up the Voyager probe project to send two spacecraft across our solar system. They were seizing a once-in-175-years opportunity when the solar system's outer planets would be in the right place for the two spacecraft to visit them all.

The probes used the gravity of Jupiter and Saturn to gain enough speed to reach Neptune and Uranus. By the time Voyager 2 reached Neptune, the probes had sent back nearly 80,000 images and five trillion bits of scientific data. Both have since traveled on to the far reaches of the solar system and they are still sending back data, although the cameras are switched off to save power and because there are no other nearby objects for them to take images.

The Voyagers' journey took them too far from the Sun to use solar energy, so they are powered by a system that creates electricity using heat from decaying plutonium. They were also the first craft with software that can independently detect problems and take corrective action.

We transmit a set of instructions to each of the probes every three months.We know how the spacecraft are doing because they are always transmitting. The data are received on the ground at one of NASA's three Deep Space Network (DSN) stations which are sited in California, Spain, and Australia. The location of these stations allows constant communication with the spacecraft as our planet rotates.

Flight director
Jefferson Hall joined the Voyager team in 1978 and became mission flight director in 1998. He and the team are still gathering valuable data for the scientists to analyze.

Today, Voyager 1's transmissions take more than 19 hours to reach Earth, while Voyager 2's take more than 16 hours. The probes' power is slowly fading, and we plan to start powering down the science instruments in 2020, although we will collect engineering data for as long as we can. Afterwards, the two probes will continue to zoom across the cosmos.

It's so exciting to be able to get all this data from interstellar space. The fact that no human-made object has ever been in this region before and may never be again in our lifetime makes it really special. It's great to see what can be done with 40-year-old technology!

"We are **delighted** to see **Voyager** still has the **capability** to acquire **unique science data**."

Voyager proof test model

Voyager 1 image of Jupiter

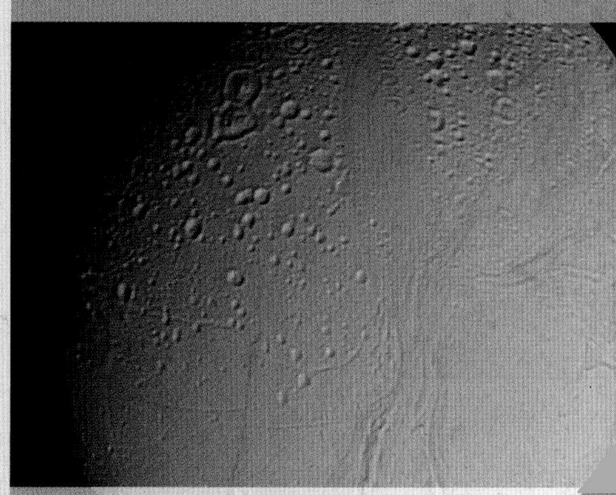

Close-up of Jupiter's moon Enceladus

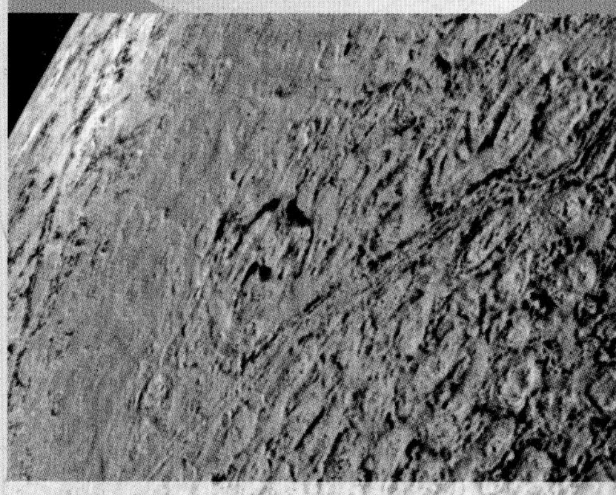

Detail of Neptune's moon Triton

DWARF PLANETS

PLANET-LIKE OBJECTS

In 2005, astronomers spotted an object almost as big as Pluto. Was Eris to be our tenth planet? No. Instead, they decided to create a new classification of dwarf planets, large objects orbiting the Sun that are not moons and have enough mass and gravity to be almost round. Five are now recognized: Pluto, Haumea, Makemake, and Eris all float on the outer edges of our solar system, while Ceres drifts in the inner Asteroid Belt.

PLUTO

WATER WORLD

Only half as wide as the US at 1,400 miles (2,380 km) across—smaller than our Moon—Pluto floats in the Kuiper Belt. It still holds three times as much water as Earth, in temperatures so low that ice-water mountains topped by freezing methane stand above frozen nitrogen plains.

Average distance from the Sun: more than 3.6 billion miles (5.8 billion km)

Visited by: New Horizons (2015)

Orbit time: 248 years

Diameter: 1,400 miles (2,300 km)

Surface temperature: –387°F (–233°C)

Discovered by: Clyde Tombaugh in February 1930

Atmosphere: none

Known moons: 5

Pluto's location was **predicted** in **1915** by US astronomer **Percival Lowell**.

Charon
This is the largest and innermost of the five known moons of Pluto. Discovered in 1978, it is thought that Pluto and Charon were formed by the same collision 4.5 billion years ago.

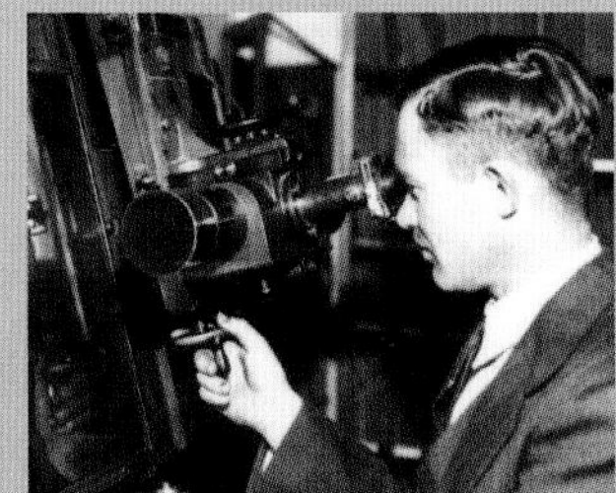

Discovery
In 1930, American astronomer Clyde Tombaugh was using a blink comparator (left) when he discovered Pluto. The machine allowed him to use his short-term memory to compare photos of the sky to see if an object had changed position.

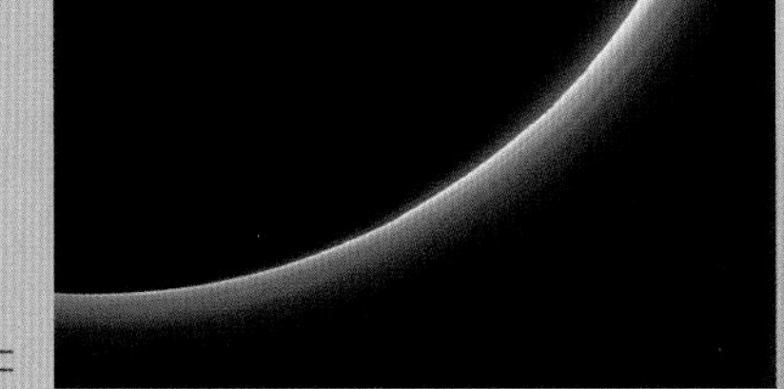

Blue haze
There is a hazy layer around Pluto which is believed to be caused by "tholins," small, sootlike particles that result from the chemical reaction of nitrogen and methane.

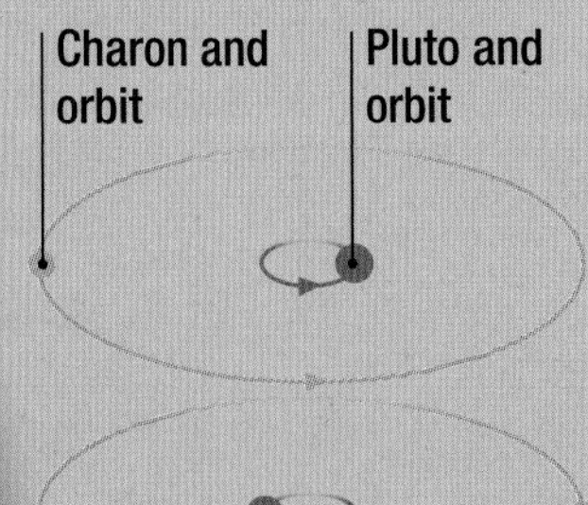

Locked together
Pluto and Charon are tied in what is described as "mutual tidal locking." This means that the same surfaces always face each other as they orbit.

Al-Idrisi Montes

Elliot crater

Young dwarf
Recent images by the New Horizons probe show a range of mountains near Pluto's equator that are younger than expected—possibly younger than anything else in the solar system. This color image enhances the different and varied features of the dwarf.

CERES

LIVING DWARF?

There may be simple, microbial life on Ceres, with its thin protective atmosphere, plenty of water, rocks, and salt deposits. Ceres floats in the Asteroid Belt between Mars and Jupiter, the only dwarf planet in the inner solar system.

Dawn on Ceres
Ceres is the closest dwarf planet to Earth and was the first to be visited, and orbited, by a spacecraft when the Dawn probe arrived there in 2015. The dwarf is covered in young craters, none of which is larger than 175 miles (280 km) across. It is possible that bright spots on the surface may be salty material bubbling up because the dwarf is geologically active.

ERIS

FROZEN DWARF

Eris is twice the distance away from the Sun than Pluto and was discovered in 2005. The dwarf planet's surface is a thin frost of nitrogen-rich ice and frozen methane.

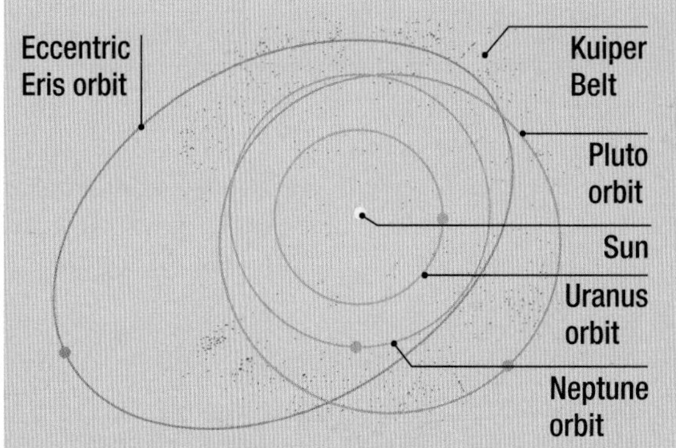

Eccentric dwarf
Eris' maximum distance from the Sun (aphelion) is 97.65 AU and its closest (perihelion) is 37.91 AU, which gives it an unusual-shaped orbit that it takes 561 years to complete. Eris has a moon, Dysnomia, and orbits in the Kuiper Belt—the outermost of the dwarf planets.

MAKEMAKE

RED DWARF

Makemake is the second brightest dwarf planet in the Kuiper Belt, after Pluto. It was first observed in 2005 by astronomers in the Palomar Observatory in California. Its reddish-brown color may be caused by a layer of frozen methane on its surface.

New moon
Astronomers recently discovered that Makemake has a small dark moon about 100 miles (160km) across. It orbits the dwarf at about 13,000 miles (21,000 km).

THE OUTER EDGE

SOLAR SYSTEM FRONTIER

Where does our solar system end? Scientists believe this point is beyond the Oort Cloud of icy debris that lies some 9.3 trillion miles (15 trillion km) away—100,000 times the distance from the Sun to the Earth. This is outside the heliosphere, the magnetic bubble of the Sun's power.

The **edge** of the **solar system** is about **9.3 trillion miles** (15 trillion km) **from the Sun**.

Moving out

From the Sun, an object traveling through space moves past the planets of the inner solar system to the outer planets and Kuiper Belt. Beyond the orbits of most of the dwarf planets lies the Oort Cloud, in which new objects are being discovered.

Orbit of Jupiter

Asteroid Belt

Orbit of Mars

Orbit of Earth

Orbit of Venus

Orbit of Mercury

Sun

INNER SOLAR SYSTEM

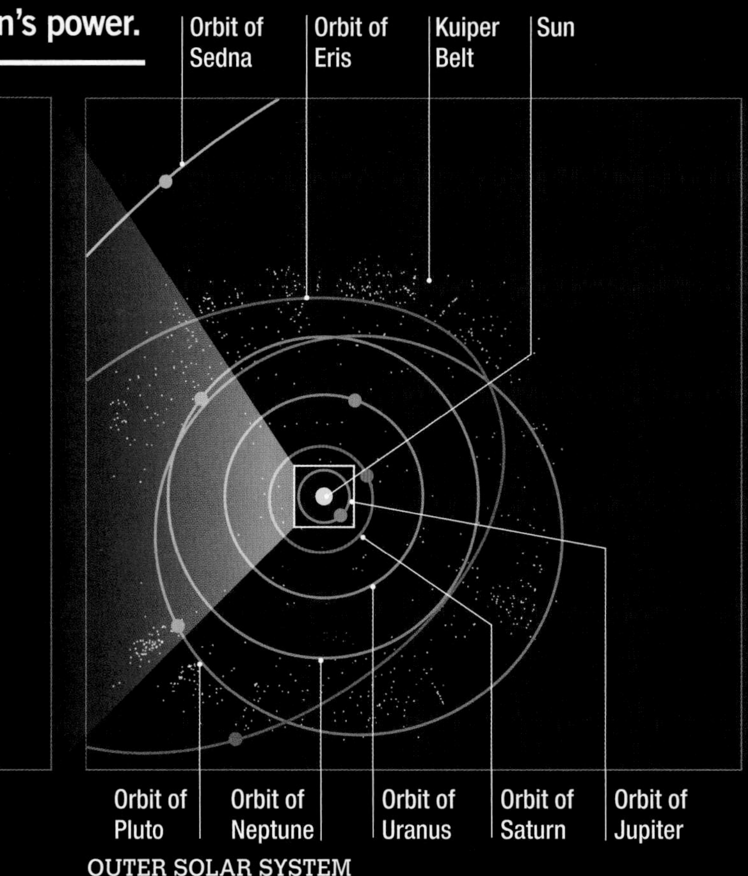

OUTER SOLAR SYSTEM

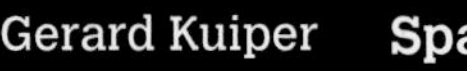

Gerard Kuiper

Jan Oort

Space tributes

Gerard Kuiper is the Dutch-American astronomer famous for discoveries and theories about our solar system. In 1951, he proposed the existence of the Kuiper Belt, which is named for him. The year before, Dutch astronomer Jan Oort had put forward the theory that comets originate from a cloud of such bodies, and the Oort Cloud was later named for him.

Long-distance traveler

More than 40 years since they were launched across our solar system, two Voyager space probes continue to communicate with Earth. In 2012, Voyager 1 became the first spacecraft to enter interstellar space and Voyager 2 is currently in the heliosheath, the outermost layer of the heliosphere.

Planet Nine

In 2016, researchers calculated that there is a possible ninth planet on the outer edges of our solar system. It is a giant object with a mass ten times that of Earth, orbiting about 20 times farther from the Sun than Neptune. No direct observations have been made.

Periodic comet Swift-Tuttle

Meteor shower from debris of Swift-Tuttle

Short-period comets

These are comets that orbit the Sun in less than 200 years, or have been seen passing near the Sun more than once. Most have been tracked to the Scattered Disk, an area beyond the edge of the Kuiper Belt.

Long-period comets

We call comets that take more than 200 years to orbit the Sun long-period comets. These are believed to have traveled from the Oort Cloud and are the most spectacular comets we see in the night sky. Long-period comets may be jolted from their orbits by a passing star or other galactic forces.

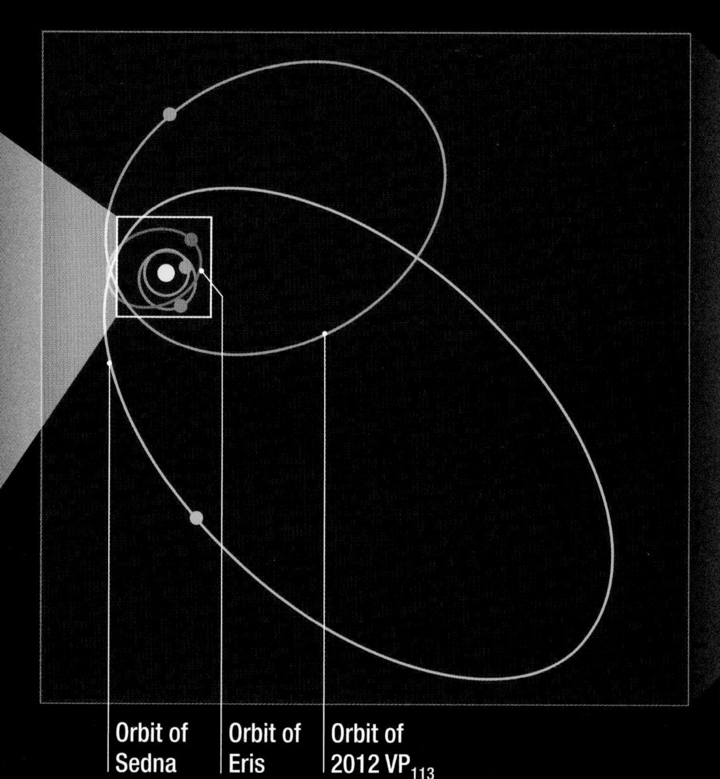

ORBIT OF DWARF PLANETS

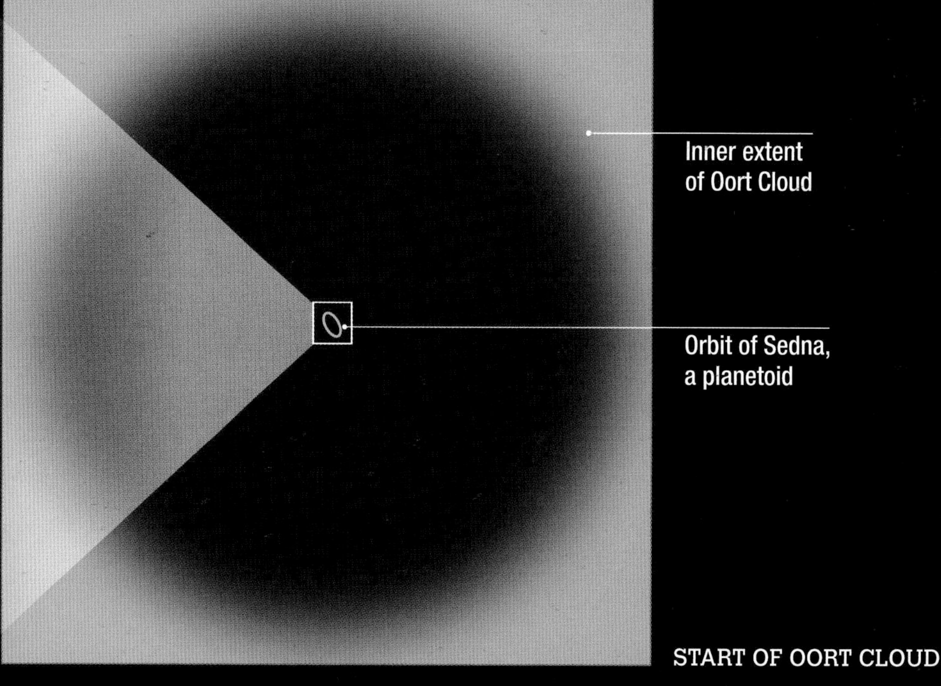

START OF OORT CLOUD

Spherical outer Oort Cloud

Heliosphere

Disk-shaped inner Oort Cloud

Kuiper Belt

Oort Cloud structure

The Oort Cloud is spread over a vast area and is thought to be the remains of the disk from which the Sun and its planets formed around 4.6 billion years ago. The region can be divided into a disk-shaped inner cloud and a spherical outer cloud, both lying beyond the heliosphere. Objects in the Oort Cloud are called "trans-Neptunian objects," but no one knows how many objects there are.

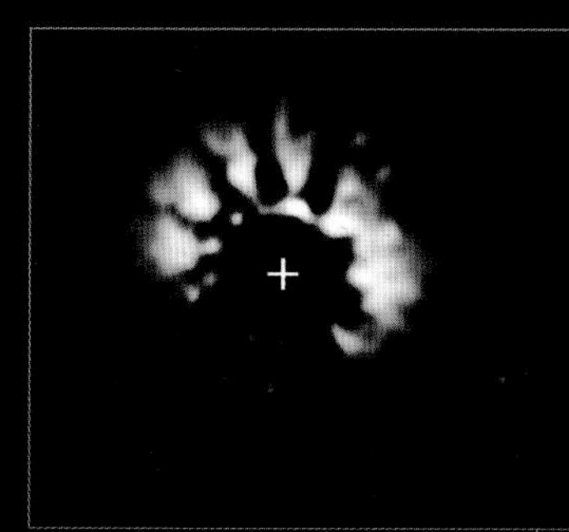

Front-on debris disk

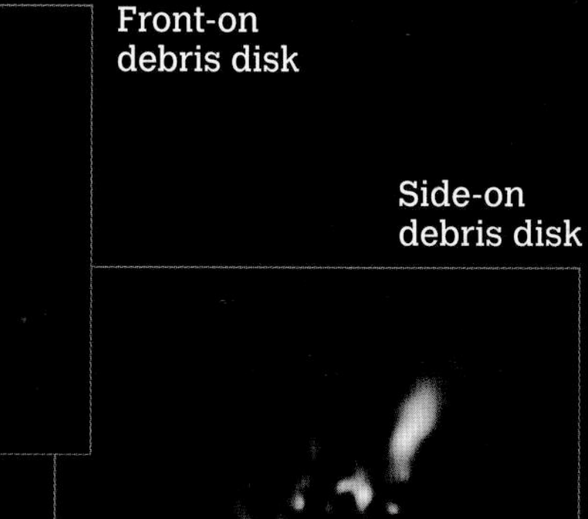

Side-on debris disk

Oort-like clouds

Hubble has taken images of five stars that have disks around them. Scientists think that the star HD 141943 (above and right) has a debris cloud exactly like our Sun had when it was being formed.

Mission to a comet

67P/Churyumov-Gerasimenko is a city-sized comet hurtling through space at up to 84,000 mph (135,000 km/h). It is a regular visitor to the inner solar system as it orbits the Sun between the paths of Jupiter and Earth every 6.5 years. The Rosetta probe chased the comet for 10 years, catching up with it in the middle of 2014 and becoming the first probe to go into orbit around a comet. It also deployed its lander, Philae, in November 2014. In September 2016, Rosetta ended its mission by deliberately crashing into the comet in a controlled impact.

SPACE

TRAVEL

SPACE RACE

Some of humanity's greatest engineering and technological feats have been achieved in the quest to explore our solar system and beyond. After World War 2, the US and the Soviet Union used captured rockets, factories, and even scientists in a "space race," and by the 1960s, people had walked on the Moon. Since then, many countries have cooperated to build the ISS, where astronauts can stay to study space.

Hasselblad Lunar Module Pilot Camera

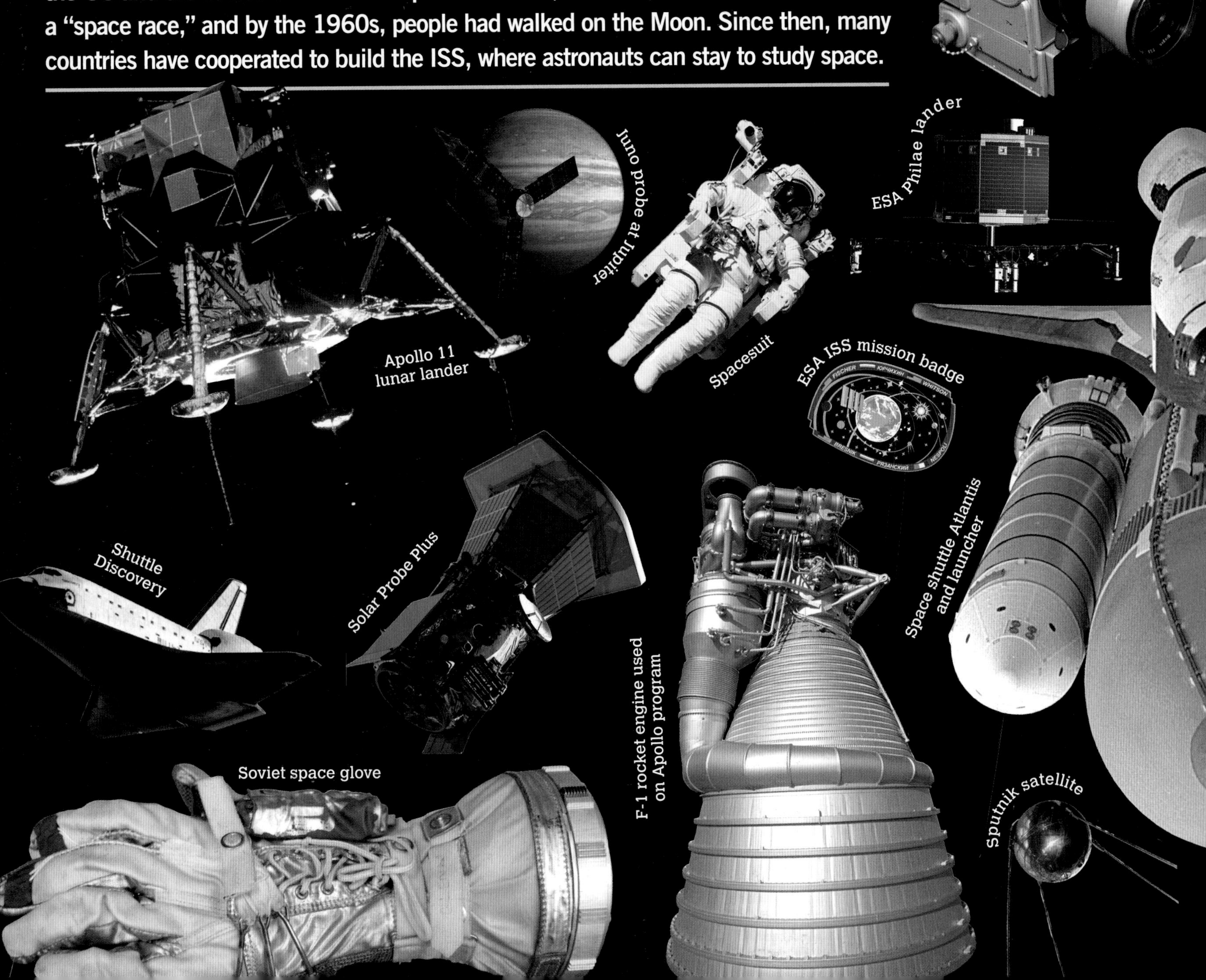

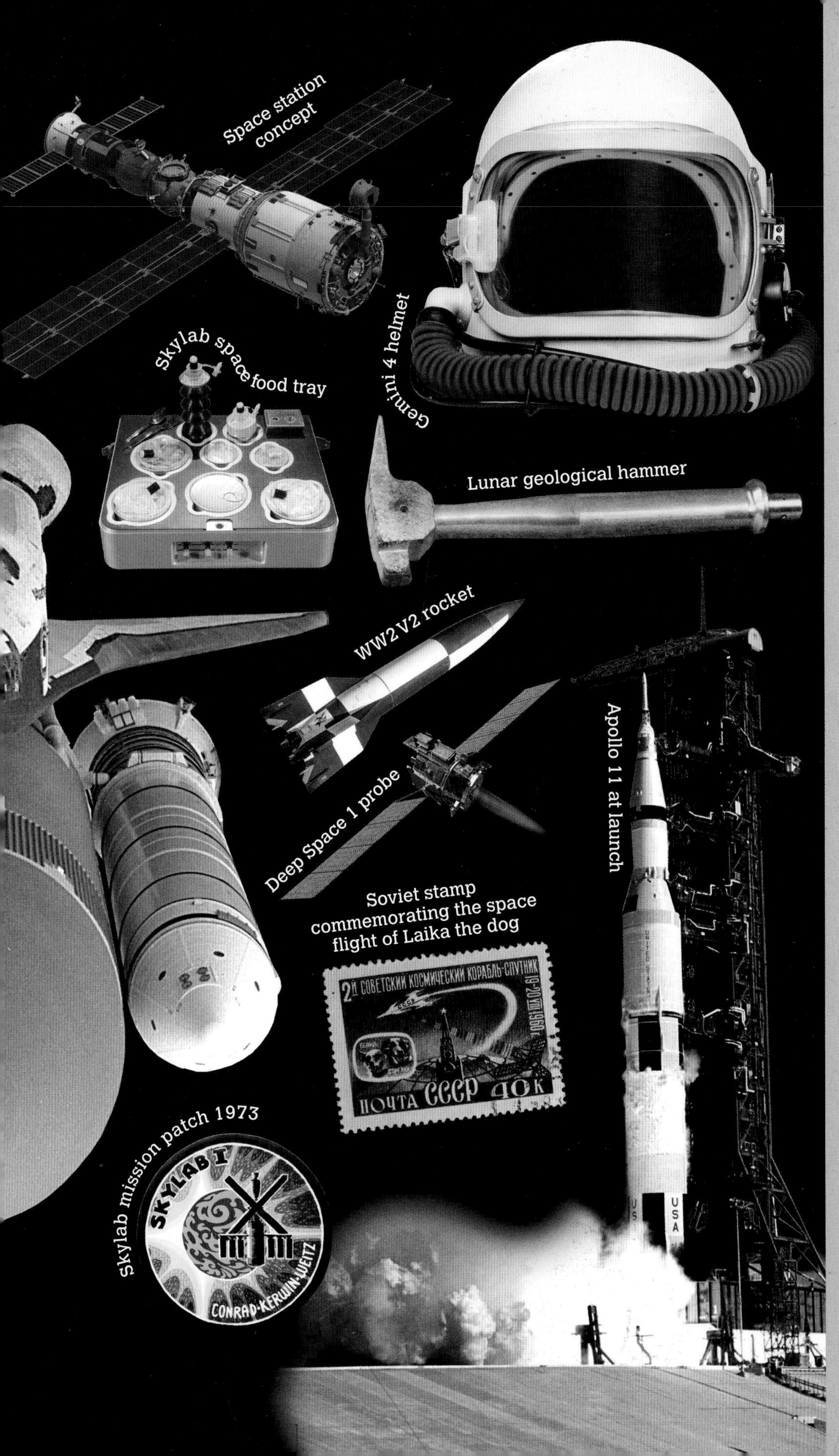

GETTING THERE FIRST

The Soviets sent the first human into space in 1961 when Yuri Gagarin was strapped in to take a jolting 1 hour, 48 minute ride around Earth. US president John F. Kennedy responded by promising that America would get a man to the Moon and back within ten years. It was a close race. Russia's Luna spacecraft took the first photographs from the surface of the Moon in 1966. Two years later, America sent the first manned spacecraft round the far side of the Moon. Then, in 1969, the Apollo 11 mission landed and Neil Armstrong's famous "one small step" won the battle.

THE NEXT GIANT LEAP

In contrast to the superpower-led space race, there were 16 nations involved in creating the International Space Station. With other space stations, the ISS has helped increase the number of space travelers so far to more than 550 people, including scientists, engineers, politicians, and even a few tourists. Space travel has become truly international. China runs the second largest fleet of spacecraft in orbit (after America), while India has launched more than 80 spacecraft. With space travel so expensive, the push into the cosmos has brought in commercial companies willing to invest the massive sums needed to get people safely into space and back again. The race has changed, but it is not over…

Apollo 13 reentry module

INTO SPACE

EARLY EFFORTS

Space officially begins at Kármán line, 62 miles (100 km) above Earth's surface. There were many early and failed attempts to send machines up through Earth's thinning atmosphere to reach outer space, but it was only after the first satellite launch in 1957 that we quickly learned how to send animals, then humans, into space.

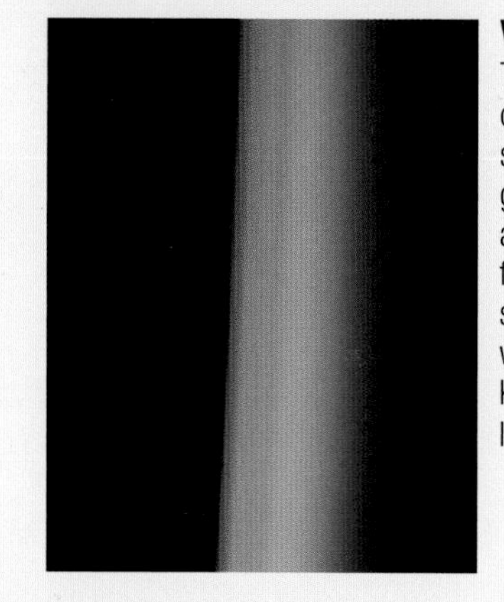

Viewed from space
This image captures the bold colors of Earth's atmosphere as seen from space. The dominant gases and particles in each layer act as prisms that scatter sunlight, filtering out certain colors, so we see mainly orange and blue. It was taken by Japanese astronaut Koichi Wakata on a visit to the International Space Station.

Earth's shield

Earth's atmosphere is a body of gases that surrounds and protects the planet. Its layers merge into each other and vary in size. Air in the atmosphere is a mixture of different gases, including nitrogen (78%) and oxygen (21%).

Mesosphere
31–53 miles (50–85 km)
The air here is too thin to breathe, and the coldest temperatures in Earth's atmosphere are found near the top of this layer.

Kármán line
62 miles (100 km)
This is commonly accepted as the boundary between Earth's atmosphere and outer space. The US defines an astronaut as someone who has flown above 50 miles (80 km).

Thermosphere
53–620 miles (85–1,000 km)
Electrically charged particles from the Sun can form aurorae in this layer, and many satellites and the ISS orbit at this level.

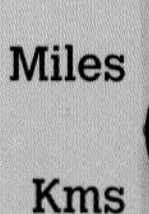

Miles: 50, 100, 150, 200
Kms: 100, 200, 300

Earth
All layer measurements are taken from ground level on Earth.

Troposphere
from ground level to 33,000 ft (10 km)
We live in this layer, air currents circulate, and nearly all weather occurs here.

Stratosphere
33,000 ft–31 miles (10–50 km)
The highest clouds are found in this layer, and this is where jet aircraft fly because it is stable.

Early rocket man

In March 1926, Dr. Robert Goddard, a New England professor, towed his rocket behind a Model A Ford truck to a launch site in the desert northwest of Roswell, New Mexico. It launched successfully, traveled at an average 60 mph (97 km/h), and reached an altitude of 41 ft (12.5 m). The flight lasted 2 seconds. By the mid-1930s, Goddard's rockets had broken the sound barrier and flown to heights of 1.7 miles (2.7 km).

Spoils of war

From 1943 onward, V2 rockets were made in Germany to crash and explode on cities such as London and Antwerp. When WW2 ended, V2 scientists and technicians were taken to the US and Russia, and in 1949, a US missile sitting atop a V2 was the first object deployed from a rocket into space.

The Huntsville Times

Man Enters Space

Soviet Officer Orbits Globe In 5-Ton Ship

Maximum Height Reached Reported As 188 Miles

'So Close, Yet So Far,' Sighs Cape

U.S. Had Hoped For Own Launch

Hobbs Admits 1944 Slaying

Praise Is Heaped On Major Gagarin

'Worker' Stands By Story

To Keep Up, U.S.A. Must Run Like Hell

Reds Deny Spacemen Have Died

Reds Win Running Lead In Race To Control Space

Front-page news

On April 12, 1961, newspapers all over the world announced that Soviet cosmonaut Yuri Gagarin was the first man in space, orbiting Earth once in a 108-minute flight in Vostok 1. The Soviets had won this part of the race.

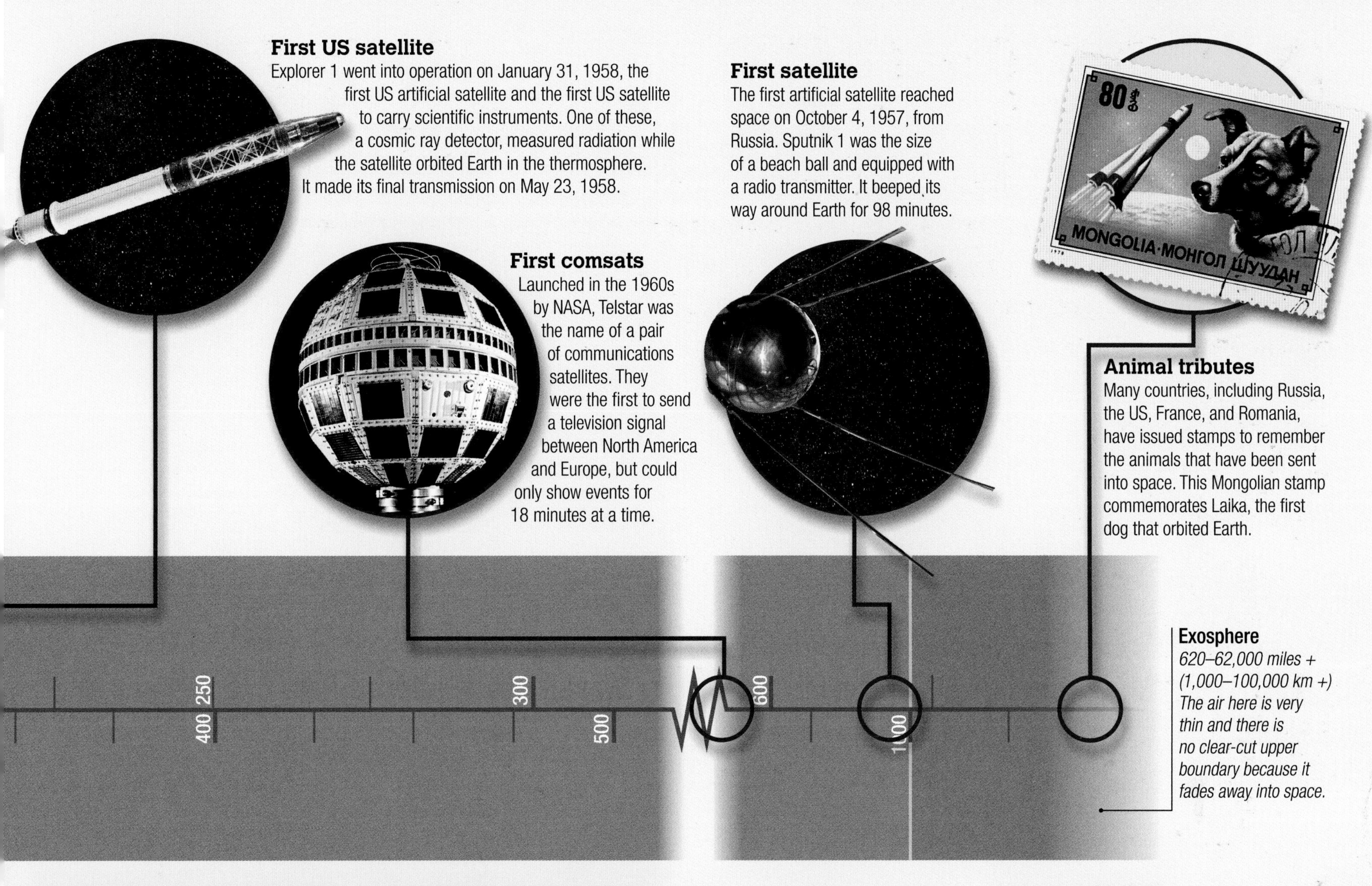

First US satellite

Explorer 1 went into operation on January 31, 1958, the first US artificial satellite and the first US satellite to carry scientific instruments. One of these, a cosmic ray detector, measured radiation while the satellite orbited Earth in the thermosphere. It made its final transmission on May 23, 1958.

First satellite

The first artificial satellite reached space on October 4, 1957, from Russia. Sputnik 1 was the size of a beach ball and equipped with a radio transmitter. It beeped its way around Earth for 98 minutes.

First comsats

Launched in the 1960s by NASA, Telstar was the name of a pair of communications satellites. They were the first to send a television signal between North America and Europe, but could only show events for 18 minutes at a time.

Animal tributes

Many countries, including Russia, the US, France, and Romania, have issued stamps to remember the animals that have been sent into space. This Mongolian stamp commemorates Laika, the first dog that orbited Earth.

Exosphere
620–62,000 miles +
(1,000–100,000 km +)
The air here is very thin and there is no clear-cut upper boundary because it fades away into space.

Earth's place in space

On April 19, 2001, the Thermal Emission Imaging System on board the Mars Odyssey took this picture of Earth and the Moon as the spacecraft headed off toward Mars. The image was taken at a distance of more than 2 million miles (3,563,735 km).

Most satellites are parked or in **orbit** around **430 miles (690 km)** above the **Equator**.

Lunar aspirations

Reaching the Moon has long been the dream of the public and scientists alike. In 1865, the French writer Jules Verne wrote *From the Earth to the Moon*, a book that inspired generations. The shape and size of the rocket in the book closely resembles the Apollo command module!

Image from *Le Voyage dans La Lune* (*A Trip to the Moon*), a film made in 1902

First image of the far side of the Moon, taken by Luna 3 in 1959

ROCKETS

SPACE TECHNOLOGY

We have gone from wondering what space is to being travelers able to power ourselves upward through Earth's atmosphere and out into space. The principles of propulsion have been developed in the 20th and 21st centuries to create rockets that can carry people and goods to space stations, to the Moon, and beyond.

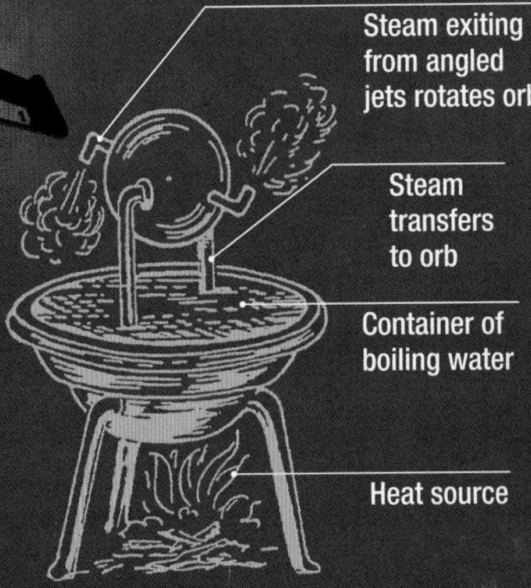

Steam exiting from angled jets rotates orb

Steam transfers to orb

Container of boiling water

Heat source

Hero's steam engine

Rockets work because exhaust gases exiting at high speed propel the rocket upward. In the first century CE, a Greek mathematician and engineer, Hero of Alexandria, invented a steam-driven machine called an aeolipile. This used the same principle to rotate a sphere.

Rocket power

The Chinese fired up the first rockets around 2,000 years ago. They experimented with gunpowder-filled tubes, developed from "fire arrows" that were used for defense. Escaping gas from the gunpowder tubes launched these first rockets.

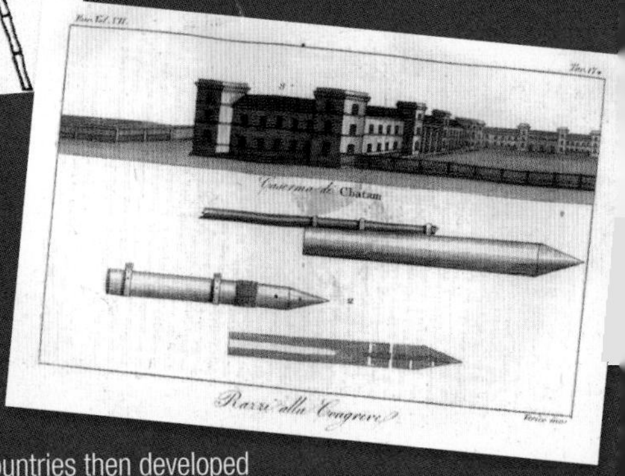

Into war

In the early 19th century, British inventor William Congreve began to experiment with rockets for military use, with iron cylinders and added warheads. Dozens of countries then developed their own rocket troops and factories. The Congreve rocket was an ancestor of the powerful rockets of today.

Blast off

The power of a rocket is needed to escape Earth's atmosphere and this happens when the thrust created by burning fuel is greater than the force of Earth's gravity bearing down on the rocket. The heavier the launch vehicle, the more fuel has to be burned to achieve the thrust needed. The influence of Earth's gravity extends a long way into space, so in order to stay in orbit around the planet, spacecraft need to go just fast enough to escape the effects of gravity but not fast enough to escape. To travel out into the solar system, large multistage rockets (top right) are needed to achieve enough thrust. They jettison each stage as fuel is used up.

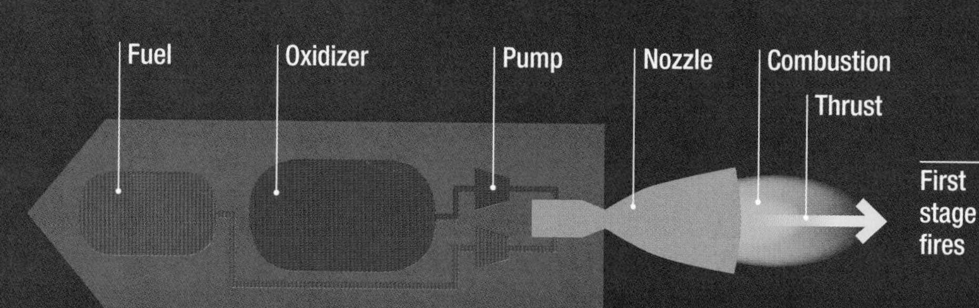

How a rocket works

A typical rocket is driven by mixing liquid oxygen (the oxidizer) and a fuel such as liquid hydrogen (the propellant). The pumps pull the cold liquids from the tanks into the engine. The ignition sparks at temperatures up to 6,000°F (3,315°C) and turns the liquids into gas, which, expelled through the nozzle, moves the rocket forward.

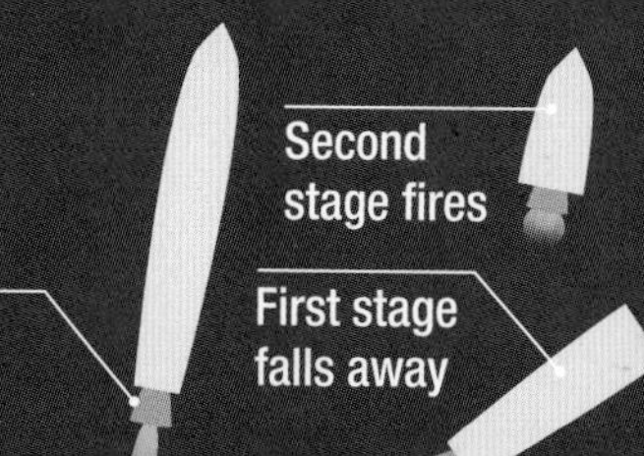

Booster stages

Multistage rockets use two or more parts, or stages. In serial staging, there is a small second-stage rocket on top of a larger first-stage rocket. After ascent, the second-stage separates, fires, and goes into orbit, while the spent first stage falls back to Earth. In parallel staging, small first stages are attached to a central second-stage rocket.

Redstone project test

Named for the Redstone Arsenal in Huntsville, Alabama, this rocket was a direct descendant of the V-2 rocket (*see p.124*). Here the rocket is at a testing site before launching on August 20, 1953. A later model carried the Mercury capsule, Freedom 7, in 1961 *(see p.136)*.

Soyuz

This series of rockets has operated since 1967. Originally developed by the Soviet Union for a lunar landing, the spacecraft was redesigned to carry payloads. Since 2011, the Soyuz space capsule is the sole transportation for crew members going to and from the ISS.

Saturn V

On July 16, 1969, a Saturn V rocket launched into space Neil Armstrong and Buzz Aldrin, the first two humans to set foot on the Moon. To this day, it is the largest rocket ever built. It needed an incredible 4,578,000 lbs (2,076,546 kg) of fuel to achieve launch.

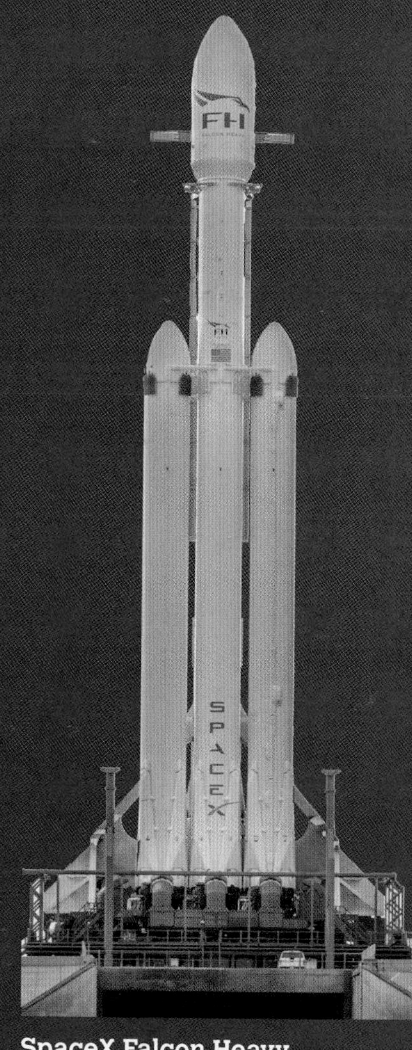

SpaceX Falcon Heavy

Rockets usually burn up in the atmosphere as they fall back to Earth. This spacecraft can carry payloads that are more than twice as heavy as any today. Add to this that its first-stage parts are reusable and you have the most powerful operational rocket in the world.

GET
MIN SEC
TBS
HR MIN SEC

Try, try, and try again

The Mission Control Center at the Lyndon B. Johnson Space Center in Houston, Texas, has managed flight control for the US human space program since 1965. This picture of the Apollo 14 mission during its docking maneuvers was taken in 1971. In the screen top right the lunar module can be seen, still attached to the third stage of the Saturn V rocket. There was a problem with the mechanism, and it took six attempts before a successful “hard dock” of the command module with the lunar module was achieved.

Stray in orbit

On August 19, 1960, The Soviet Union launched Sputnik 5 into space. On board were 2 stray dogs from the streets of Moscow, 40 mice, 2 rats, 1 rabbit, several fruit flies, and some plants. This is Belka, which means "squirrel," one of the two dogs on board of only the second spacecraft to carry animals into orbit. They orbited for one day and became the first animals to return to Earth safely.

ANIMALS IN SPACE

Many creatures were used to test out the safety of space travel before scientists were willing to risk a human life in the dangerous world of the space capsule. In 1947, fruit flies became the first animals—and fastest insects on (above) the planet—when they soared 68 miles (110 km) up inside a rocket. The practice continues. In 2017, worms on the ISS helped study the effects of space travel on cell activity.

Space monkeys

Many monkeys have traveled into space, including Albert II, a Rhesus monkey that was the first monkey and mammal in space on June 14, 1949. It was not until the 1960s, when it was established that humans could survive the rigors of spaceflight, that the number of monkeys and apes being sent into space was reduced in favor of smaller animals.

Publicity shot of Baker the squirrel monkey

Mutlik the macaque took part in a joint US/Russian/French flight

Rhesus monkey Sam before the flight on a Little Joe rocket

Space menagerie

A whole host of different animals have made the journey into space on rockets and shuttles. Among other things on the ISS, astronauts are investigating whether animals can cope with microgravity and if bees can make honey in space.

Arabella the spider weaving the first space web

Bullfrogs

Steppe tortoises

Guinea pigs

Mice

Rats

Honey bees

Fruit flies

Dogs in space

The Soviets believed that dogs were the perfect test subjects and sent more than 20 of them into space. The scientists thought that they would be able to put up with long periods of being inactive. The dogs that took part were seen as heroes, and remembered with stamps and monuments.

Lighthearted ad for a burger to celebrate the flight of Sputnik 2 carrying Laika the dog

Malyshka flew low-orbital flights in 1951

Tough critters

Tardigrades, also known as water bears, are tiny 1-mm long creatures that live in droplets of water.
For 12 days in September 2007, 3,000 tardigrades traveled into space on the outside of ESA's orbiting Foton-M3 mission and survived.

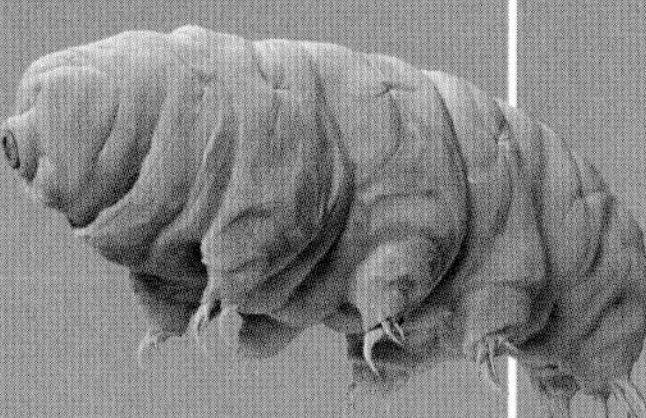

Tardigrade

Laika the top dog

On November 3, 1957, Sputnik 2 was launched with a little dog, Laika, on board. This was only the second time a spacecraft had been launched into Earth's orbit, and the first time a living creature had been on board. Laika was a stray, picked up from a Moscow street only a week before launch. She was not the first animal in space, but she was the first to orbit Earth, and provided data on how living organisms reacted to launch and conditions in space.

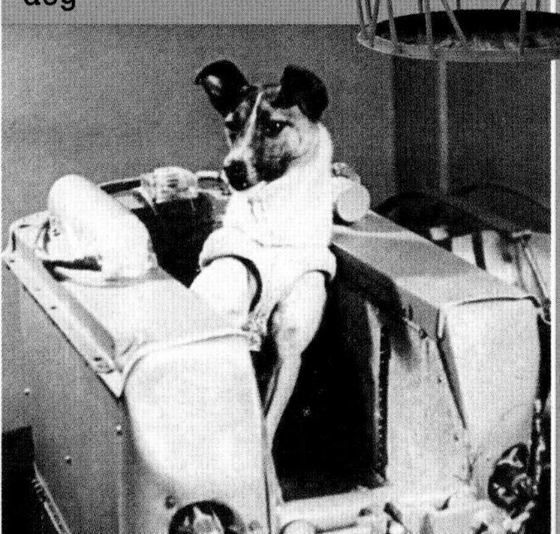

Sputnik 2

Laika the dog

REACHING FOR THE MOON

Only ten years passed between the first spacecraft landing on—or rather, smashing into—the Moon and human footprints making their imprint on lunar soil. During that time, the first human was blasted into space, probes tested orbit patterns and hunted down smooth areas of the Moon's cratered surface for a safe landing. In 1969, this chapter of the story of space exploration ended triumphantly with Neil Armstrong's "giant leap."

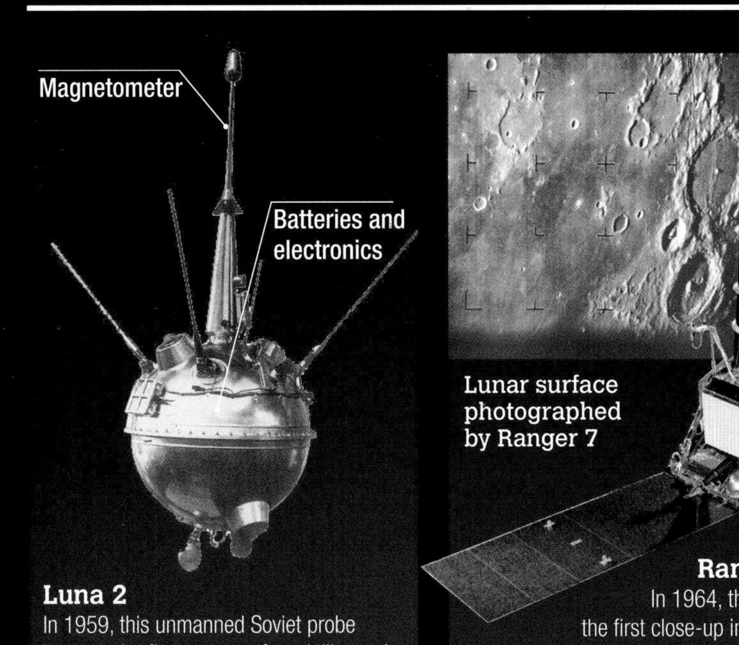

Luna 2
In 1959, this unmanned Soviet probe became the first spacecraft to deliberately crash into the surface of the Moon. The mission confirmed that the Moon did not have a magnetic field, and that there was no evidence of radiation belts.

Lunar surface photographed by Ranger 7

Ranger 7 probe

Ranger 7
In 1964, this US space probe sent back the first close-up images of the lunar surface. This information helped NASA to plan for the Apollo program and landing their astronauts on the Moon's surface from 1969.

Covered capsule

Luna 9 probe

Photograph of the lunar surface

Luna 9
Two years before the US flew astronauts to the Moon, this Soviet probe landed there, ejecting a landing capsule just before impact. The robotic lunar station was about 2 ft (0.6 m) across and weighed around 218 lbs (99 kg). It took four panoramic views of the surface.

Barren target
The probes and spacecraft that visited the Moon found it covered with impact craters, extinct volcanoes, and wide plains caused by ancient lava flows. Three Luna missions brought back samples and six Apollo missions carried back 842 lbs (382 kg) of rocks and soil. Among other things, these have revealed that the Moon probably formed from debris when Earth, a short time after its own formation, was hit by an object around the size of Mars.

The **Moon orbits** Earth every **27.3 days**.

Detail of probe foot in lunar soil

Lunar surface

Solar panel

TV camera

Surveyor 1 probe

Surveyor 1

On June 2, 1966, the US landed its first spacecraft, Surveyor 1, on the Moon. It had a soft landing in the Ocean of Storms and just 30 minutes later began to transmit images to Earth. Contact with the probe was lost in January 1967.

Lunar lander above the Moon, with the Earth visible

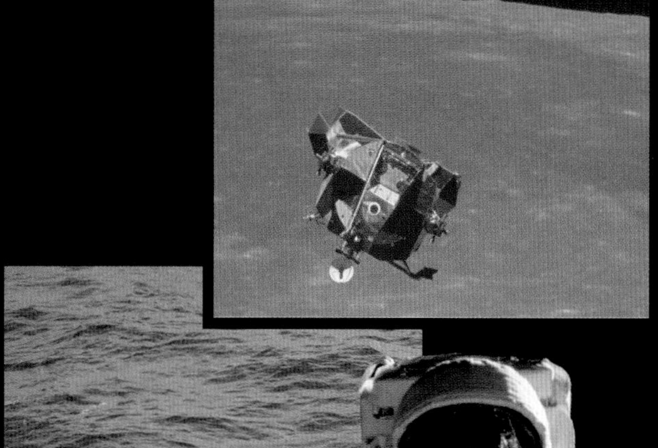

Crew being recovered in the Pacific Ocean

Astronaut Buzz Aldrin on the lunar surface

Apollo 11

On July 20, 1969, Neil Armstrong's words, "That's one small step for [a] man, one giant leap for mankind," were broadcast from the Moon. He and Buzz Aldrin spent 21 hours there, before lifting off in part of the lunar module Eagle to rejoin pilot Michael Collins in Apollo 11 for the return to Earth.

Geological sampling and recording

Spacewalk to retrieve film from external cameras

Third use of the lunar rover

Apollo 17

The sixth and last successful Apollo mission to land people on the Moon took place in 1972. The lunar rover was used to survey and collect samples, as well as to carry scientific equipment for experiments. NASA had turned its attention to building a space station, Skylab. Astronaut Eugene Cernan became the last astronaut to set foot on the Moon.

Walking in space

The first person to go on a spacewalk or extravehicular activity (EVA) was the Russian cosmonaut Alexei Leonov on March 18, 1965. The spacewalk lasted for 10 minutes. Today, astronauts can spend up to eight hours at a time on a spacewalk. Here, astronaut Mike Hopkins is on the second of two spacewalks spread over a four-day period in 2013. Together with Rick Mastracchio, whose image can be seen in Hopkins' helmet visor, he is changing a degraded pump module on the outside of the International Space Station.

SPACESUITS

SPACE TECHNOLOGY

Spacesuits are mini-spacecraft. They are puffed up with oxygen for the astronauts to breathe, and release the harmful carbon dioxide they exhale. They keep out the freezing cold of space, the scorching heat of direct sunlight, and flying space dust. Over time, they have become less cumbersome and easier to wear for longer periods of time.

Dangers of space

Backpacks offer astronauts life-support systems, and can be rocket-powered so they can float in space. Temperatures can be extreme, and astronauts may spend up to eight hours on a mission outside a spacecraft, so it is vital that suits are tested thoroughly.

Sealed for survival on spacewalks

Testing suits for leakage in water tanks

Evolution from flight

Spacesuits developed from clothing worn by top pilots in the 1950s to protect them if cockpit pressure was lost in the cold, thin air high up in the atmosphere above 50,000 ft (15,250 m). The suits had to be even tougher to cope with the astronauts' needs in space. The redesigned clothing was filled with oxygen, and carried a water supply, and even a toilet.

Pilot Joe Walker in 1958, in an early partial pressure suit to protect him in his X-1E plane at high altitude

The crew of the X-15, a rocket-powered aircraft operating from 1959

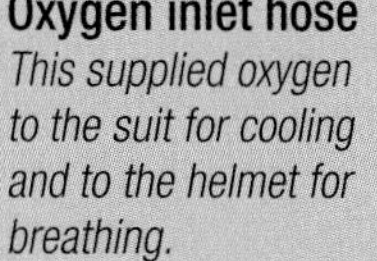

Visor seal
An airtight seal ensured that air reached the astronaut via the padded helmet with its pull-down visor.

Sealing ring
This joined the helmet to the top of the spacesuit.

Oxygen inlet hose
This supplied oxygen to the suit for cooling and to the helmet for breathing.

Gloves
The gloves attached with a ring connector and were held in place with a flap of fabric fixed to the suit by a zipper.

Footwear
Loose socks of airtight fabric were attached directly to the legs and the boots laced up tightly on top to keep the socks from ballooning.

Silver suited

This spacesuit was used during the early days of NASA's Project Mercury program (1958–1963). The outer part of the suit was made with green nylon coated with aluminum powder, which gave it the famous silver coloring that reflected heat.

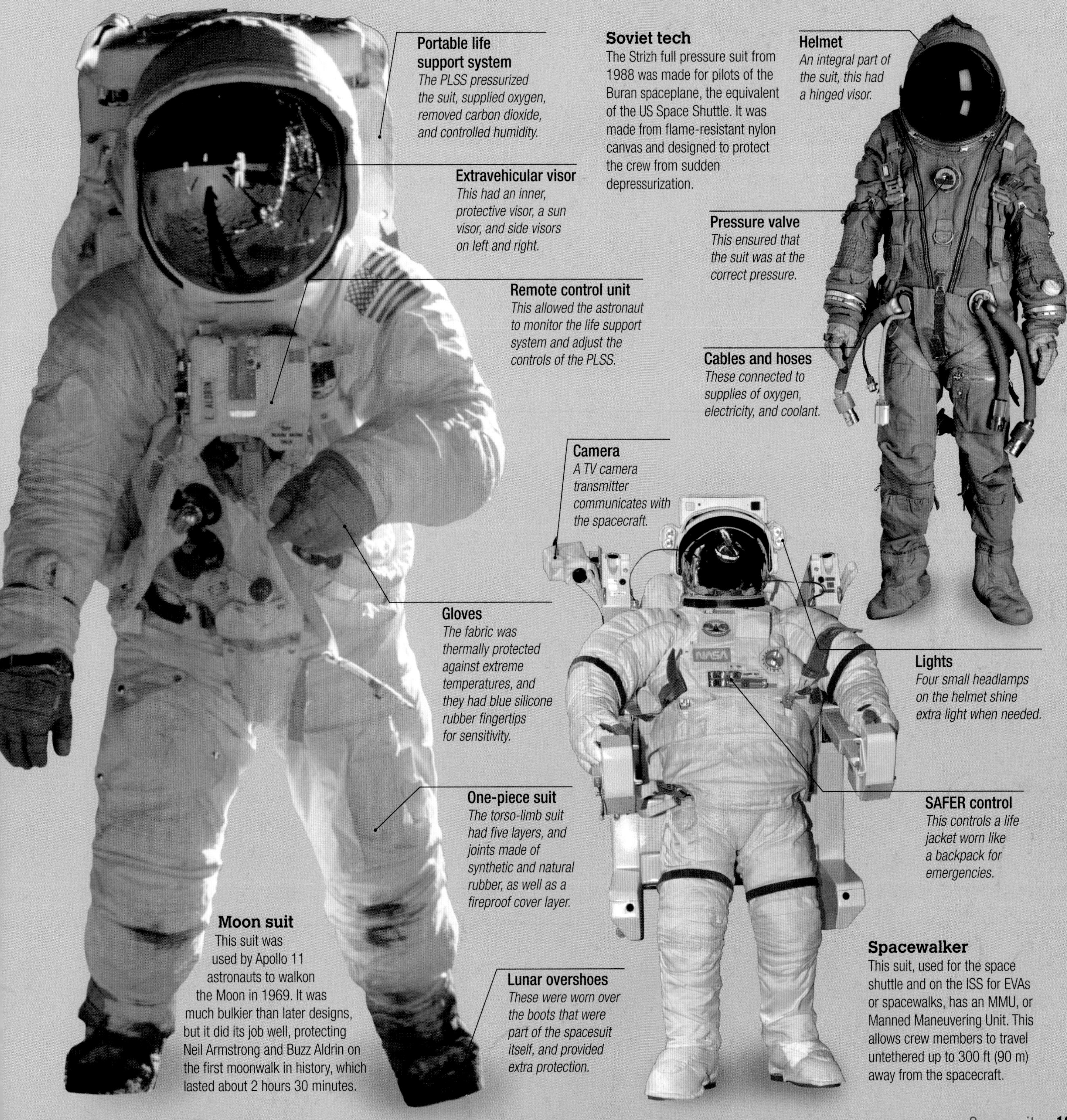

Soviet tech
The Strizh full pressure suit from 1988 was made for pilots of the Buran spaceplane, the equivalent of the US Space Shuttle. It was made from flame-resistant nylon canvas and designed to protect the crew from sudden depressurization.

Moon suit
This suit was used by Apollo 11 astronauts to walkon the Moon in 1969. It was much bulkier than later designs, but it did its job well, protecting Neil Armstrong and Buzz Aldrin on the first moonwalk in history, which lasted about 2 hours 30 minutes.

Spacewalker
This suit, used for the space shuttle and on the ISS for EVAs or spacewalks, has an MMU, or Manned Maneuvering Unit. This allows crew members to travel untethered up to 300 ft (90 m) away from the spacecraft.

MOON LANDING

NATURAL SATELLITE EXPLORATION

Apollo 11 and crew take off in Saturn V

When, in 1969, Neil Armstrong stepped off the lunar module ladder to kick up the first cloud of moon dust, the world was watching. This was a landmark moment in space exploration and in human history—our species had reached an alien world. A TV audience of 600 million, one in every five people on Earth at the time, marveled at the blurry images. There really were people standing on that silvery disk in the sky!

July 16
Launch takes place from the Kennedy Space Center.

July 19
Spacecraft out of contact behind Moon.

Orbit
Spacecraft orbits Earth while crew does checks.

Landing
Eagle lands on the Moon.

Orbiting
Command module continues to orbit.

Meanwhile . . .
Strapped to their seats in the command module on the top of Saturn V were three astronauts, Neil Armstrong, Buzz Aldrin, and Michael Collins. After one and a half orbits of Earth they headed for the Moon.

S band antenna
This was for a unified tracking and communications system.

RCS thruster assembly
One of the eight thrusters used for takeoff and ascent.

Housing
Contained the aerozine 50 fuel needed to leave the Moon.

Descent stage
This is the lower half of the module.

Landing legs
These unfolded and locked for landing.

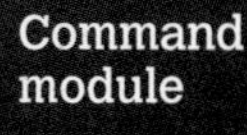
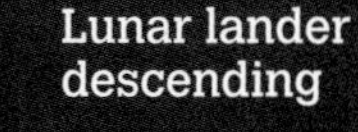
Command module

Lunar lander descending

Arrival
Three days later, the spacecraft was in lunar orbit, and a day after that, Armstrong and Aldrin climbed into the lunar module Eagle for the descent to the Moon's surface. Collins stayed in the command module and continued to orbit.

"The Eagle has landed"
Eagle touched down at 20.18 UTC on July 20, with only 30 seconds of fuel left. At 02.56 UTC on July 21, with more than 600 million people watching on television back on Earth, Armstrong became the first human to stand on the Moon.

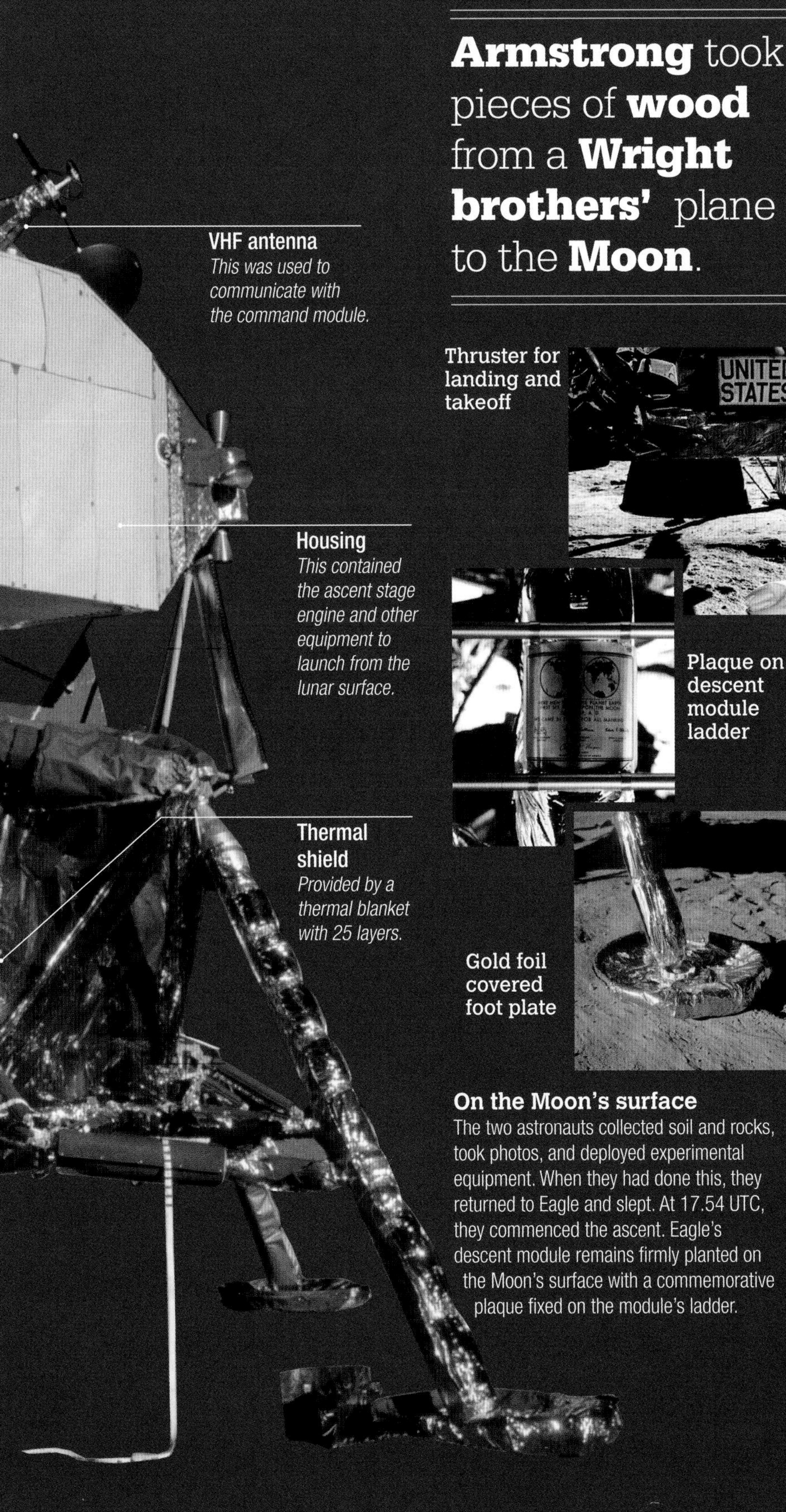

Armstrong took pieces of **wood** from a **Wright brothers'** plane to the **Moon**.

Thruster for landing and takeoff

Plaque on descent module ladder

Gold foil covered foot plate

On the Moon's surface

The two astronauts collected soil and rocks, took photos, and deployed experimental equipment. When they had done this, they returned to Eagle and slept. At 17.54 UTC, they commenced the ascent. Eagle's descent module remains firmly planted on the Moon's surface with a commemorative plaque fixed on the module's ladder.

Other visitors

Since this historic landing, another ten astronauts have walked on the Moon. Together with those who have orbited the Moon, the total so far is 24 people who have visited our satellite. However, several unmanned rovers have landed successfully and sent back information, including China's Chang'e-3 in 2013.

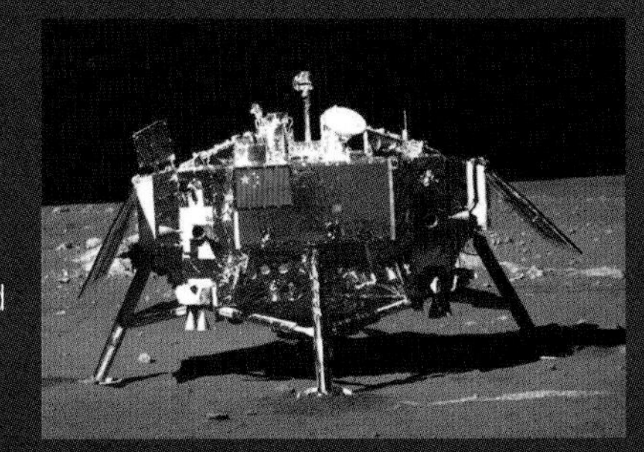

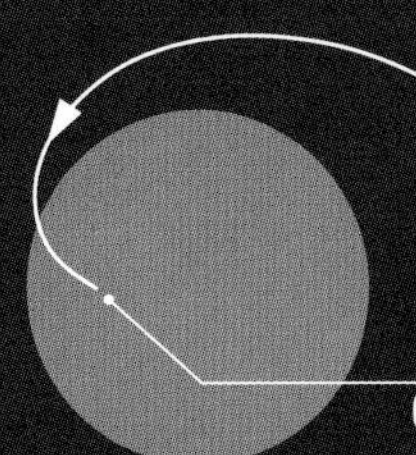

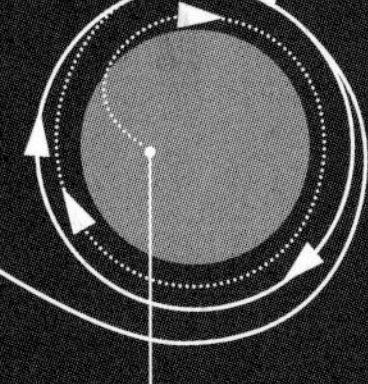

Setting off

Buzz Aldrin accidentally broke the ignition switch. Without this, they could not activate the ascent engines. They were stuck—until Buzz jammed a pen into the switch and the engines fired.

Launch pad
This was the descent stage of the lander.

Ascent stage
This is fired up for rendezvous with the command module.

Separation
The ascent module is jettisoned.

Command module
This is oriented for reentry.

Takeoff
The ascent module leaves the Moon.

Heading home

The ascent module docked with the command module, and the crew and materials transferred before the Eagle was jettisoned. Then the command module headed for Earth. Entering Earth's atmosphere, it shed flaming pieces of covering, as it was supposed to, to protect the interior.

Pickup

Splashdown was in the Pacific Ocean southwest of Hawaii, only a few miles from the U.S.S. Hornet, the recovery ship. The crew had to stay in quarantine for 21 days to avoid any contamination.

Exploring the Moon

When Apollo 15's lunar module, Falcon (left), arrived on the smooth plain of the Hadley Delta, the commander, David Scott, announced, "We're sure in a fine place here." For the first time, three seven-hour long moonwalks could be achieved, and a lunar rover would allow the crew to travel a greater distance from the module and collect more material. Here, the lunar module pilot Jim Irwin is loading up the lunar rover with tools and equipment for the first trip.

LUNAR ROVERS

SPACE TECHNOLOGY

Three Apollo missions—15 (1971), and 16 and 17 (1972)—gave scientists huge amounts of information about the Moon because they each carried a rover, an electric car that allowed astronauts to cover more ground. Russia had already landed a rover, Lunokhod, in 1970 and China followed in 2013 with the Yutu vehicle that has discovered a new kind of lunar rock.

Apollo 17 mission commander Eugene Cernan in the lunar roving vehicle on its first EVA

Tire tracks made by NASA's Lunar Reconnaissance Orbiter around the Apollo 17 mission site

Landing procedure

The folded rover was anchored to one central strut of the lander's body. It was held there by locking pins, cables, shock absorbers, and other equipment that meant that one astronaut could lower it to the surface on their own if necessary. The first thing astronauts did once the rover was fully unpacked was test out the controls.

Moon buggy

The battery-powered, four-wheel drive lunar rover is popularly known as the Moon buggy because of its resemblance to a dune buggy. And the rover had to negotiate terrain that was not unlike the irregular surfaces found among many dunes near Earth's seas.

Apollo LRVs

Lunar rovers were taken to the Moon's surface and left there on the three Apollo missions. On the Apollo 17 mission, the rover drove a total of 22.3 miles (35.9 km), reaching a top speed of 11.2 mph (18 km/h). The scientific return from these missions has proved invaluable, with far more material brought back to Earth than was possible previously.

High-gain antenna
The high-gain antenna was used to transmit pictures and data, but did not function when the rover was on the move.

Low-gain antenna
This maintained audio communication, even when the rover was moving around.

Color television camera
Controlled remotely from Earth, this allowed panoramic filming of astronauts carrying out tasks.

Control and display console
This controlled the motors that made the rover move, steer, and brake. The console also had an odometer to show distance and bearing from the lunar module.

4-wheel steering
The rover could negotiate rough ground and execute a U-turn within a 10-ft (3-m) radius.

Aluminum frame
This consisted of 2,219 welded tube assemblies in a three-part chassis, hinged in the center so it could be folded neatly.

Seatbelt fixing
The seatbelts hooked together with Velcro and stopped the astronauts being bumped off by the rough terrain.

Storage
There was storage both under and behind the seats for rock-sampling gear. There were special sample bags for storing rocks and soil.

The **Apollo lunar roving vehicles** and **test models** were built by the **Boeing company** at a cost of **$38 million**. In today's money, that **equates** roughly to **$235 million**.

16mm film camera
All cameras, film magazines, and lenses had black anodized surfaces to eliminate reflections.

Stereo camera
These cameras simulate human binocular vision, capturing three-dimensional images.

Surface Electrical Properties antenna
One of the experiments carried out used radio waves to "see" down under the surface looking for boulders and water.

Wire mesh tires
Using piano wire mesh considerably reduced the weight of the rover's sizeable tires.

Batteries
The rover was driven by two non-rechargeable 36-volt silver-zinc potassium hydroxide batteries, which powered the drive and steering motors.

Lunokhod 1

Yutu

Solar panel "wings"
The solar panels have kept Yutu operating despite the fact it lost mobility after a month.

Other rovers

Three countries have had rovers on the Moon—the Soviet Union, the US, and China. In 1970, the Soviet Union landed the unmanned Lunokhod 1. There followed the last three Apollo missions during 1971 and 1972. Lunokhod 2 arrived there in 1973. The Chinese vehicle Yutu (Jade Rabbit) landed in December 2013 and is still gathering data.

NASA Ames K10 rover

Lunar Electric Rover for future missions to the Moon

Future rovers

NASA developed the Lunar Electric Rover for future trips to the Moon. It is 10 ft (3 m) tall, and has a pressurized interior. Two "suitports," or hatches, at the back will give astronauts space to get ready so they can exit in only 15 minutes. However, NASA has Mars in its sights now (*see pp.194–195*), so there are no immediate plans for its use.

Gold and space

Gold came to Earth from space carried by meteorites and now we are returning some of it. The shiny, precious metal is used to thinly coat the visors of astronauts to protect them from the dangerous effects of solar radiation. Spacecraft and satellite electronics are protected from corrosion from ultraviolet light and X-rays, and gold makes excellent electrical contacts in onboard electronics.

INVENTIONS

Space exploration has led to many inventions that improve life on Earth. You might expect rocket travel to improve wireless headsets or scratch-resistant lenses. Maybe it is not a shock that water filters and firefighter suits have benefited, or that the memory foam in your mattress was developed by NASA. But did you know modern ear thermometers use infrared technology that measures the temperature of stars?

Gold on electrical connections

Gold

Gold is a useful metal because it does not tarnish. Experiments in space have led to several inventions on Earth (above and right). And the same process that gold-plates the Oscars will be used on the James Webb Space Telescope to reflect light in infrared wavelengths, making it easier to spot distant stars.

Gold coating infrared sensors on in-ear thermometers

Gold in high-speed circuitry

Thermal image of a house showing heat loss

Temperature control

Radiant barrier insulation was developed to allow astronauts to work in Apollo modules in shirt-sleeves. The insulation is now used in many newly constructed domestic buildings to conserve up to 20% of energy in the winter.

Tough coating

One challenge for NASA was to protect their rockets and gantries from rust at launch sites on Earth. In the early 1980s, a coating was made that bonds with steel and dries quickly to a hard finish. In the mid-1980s, 225 gallons (852 liters) were applied to protect the inside of the Statue of Liberty.

Statue of Liberty

Structural support

Today, there are more than 550 buildings and bridges in the earthquake-prone areas of San Francisco, Tokyo, and Taiwan that are fitted with seismic dampers (shock isolation systems) developed from technology used to solve a problem for NASA. When a rocket is on the launch pad, large swing arms deliver fuel and power. The dampers control the shock of takeoff, making sure that the swing arms are guided back into their cradles during launch. These same dampers are integrated in buildings to resist earthquake shocks.

Many architects of Tokyo skyscrapers integrate seismic dampers in their structures.

Catching the moment

During the 1990s, a NASA team was working to create cameras that would satisfy the need for scientific detail but were also small enough for interplanetary travel. They came up with the CMOS (complementary metal-oxide semiconductor) image sensors. These were developed in the 2000s as a camera to fit in slim cellphones.

Smartphone technology from space

Whenever you **take a picture** with your **smart phone**, you are using **NASA technology**.

Versatile materials

New tensile fabrics made from fiberglass and Teflon were originally designed for spacesuits in the 1970s. Today they are used as roofing material in buildings that include stadiums, pavilions, malls, and airports.

Dallas Cowboys' stadium with its tensile fabric roof

"Space" blankets

The shiny, reflective material that protected all the Apollo flights is now used back on the ground. The thermal foil is made into blankets used by runners to keep warm and distributed by rescue agencies in emergencies.

Thermal foil blankets

SPACE STATIONS

SPACE TECHNOLOGY

Space stations were the next step in helping us to understand the effects of long periods of weightlessness and how astronauts might cope with long journeys. They are giant satellites filled with oxygen so that people can live in some comfort. They allow scientists to study life in space and are perfectly placed to look out into the Universe with no atmosphere to block the view—and to look back and study Earth.

Salyut 1, launched in 1971

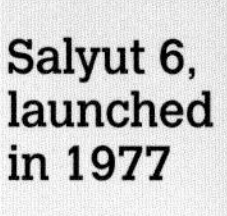

Salyut 6, launched in 1977

Early Soviet steps

The first space station program was a series of seven Soviet space stations launched between 1971 and 1982. They were used as science laboratories or military reconnaissance platforms, and six of them were crewed. The long-duration stay crews of the Salyuts 6 and 7 hosted international guest astronauts for short stays.

Scale model of an inflatable circular space station

Mir spent **15 years** in Earth **orbit**, three times its **planned lifetime**.

1977 drawing of a "spider" concept station

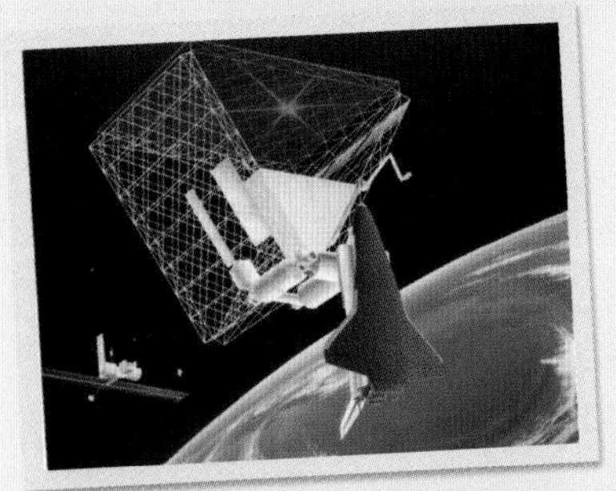

1984 solar-panel "roof" concept

High flying ideas

People have imagined what it would be like to live in space for more than 100 years. In October 1869, readers of *Atlantic Monthly* began reading Edward Everett Hale's story "The Brick Moon," which told of a sphere orbiting Earth and helping ships navigate.

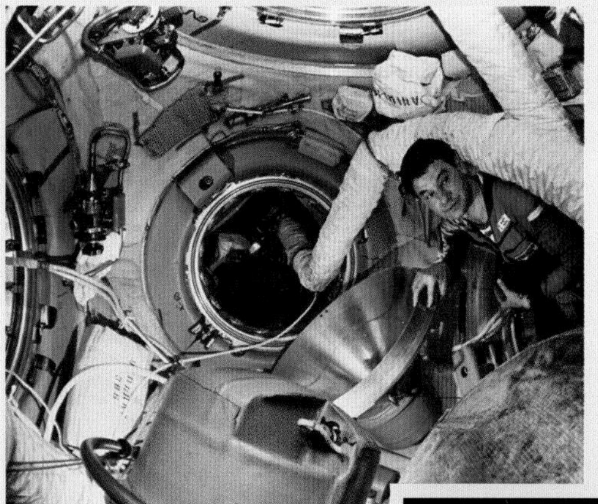

Cosmonaut in central node connecting modules

Built: 1986–1996

Launch sites: Baikonur Cosmodrome and Kennedy Space Center

Fully crewed: 3

Length: 62.3 ft (19 m)

Width: 101.7 ft (31 m)

Mass: 285,940 lbs (129,700 kg)

Orbital period: 91.9 mins

Status: deorbited March 23, 2001

Mir with the space shuttle Atlantis in dock

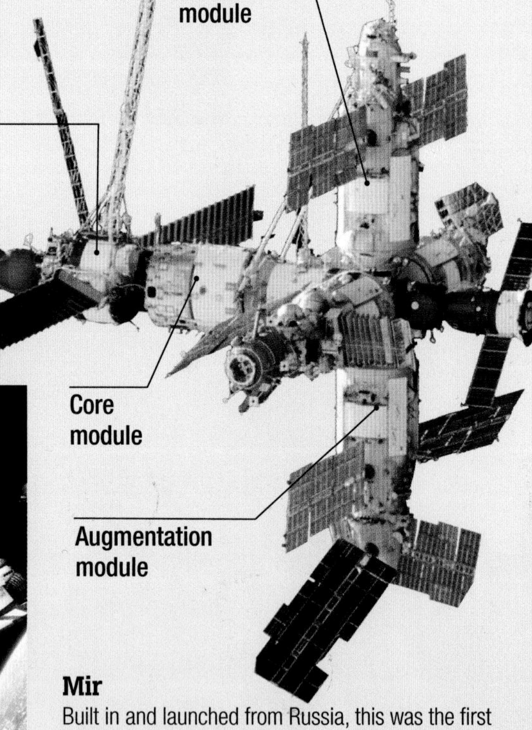

Mir

Built in and launched from Russia, this was the first major international station. From 1986 to 2001, Mir was visited by 125 cosmonauts and astronauts from many countries and hosted more than 23,000 scientific experiments. The longest consecutive spaceflight was on Mir—Valeri Polyakov spent nearly 438 days there.

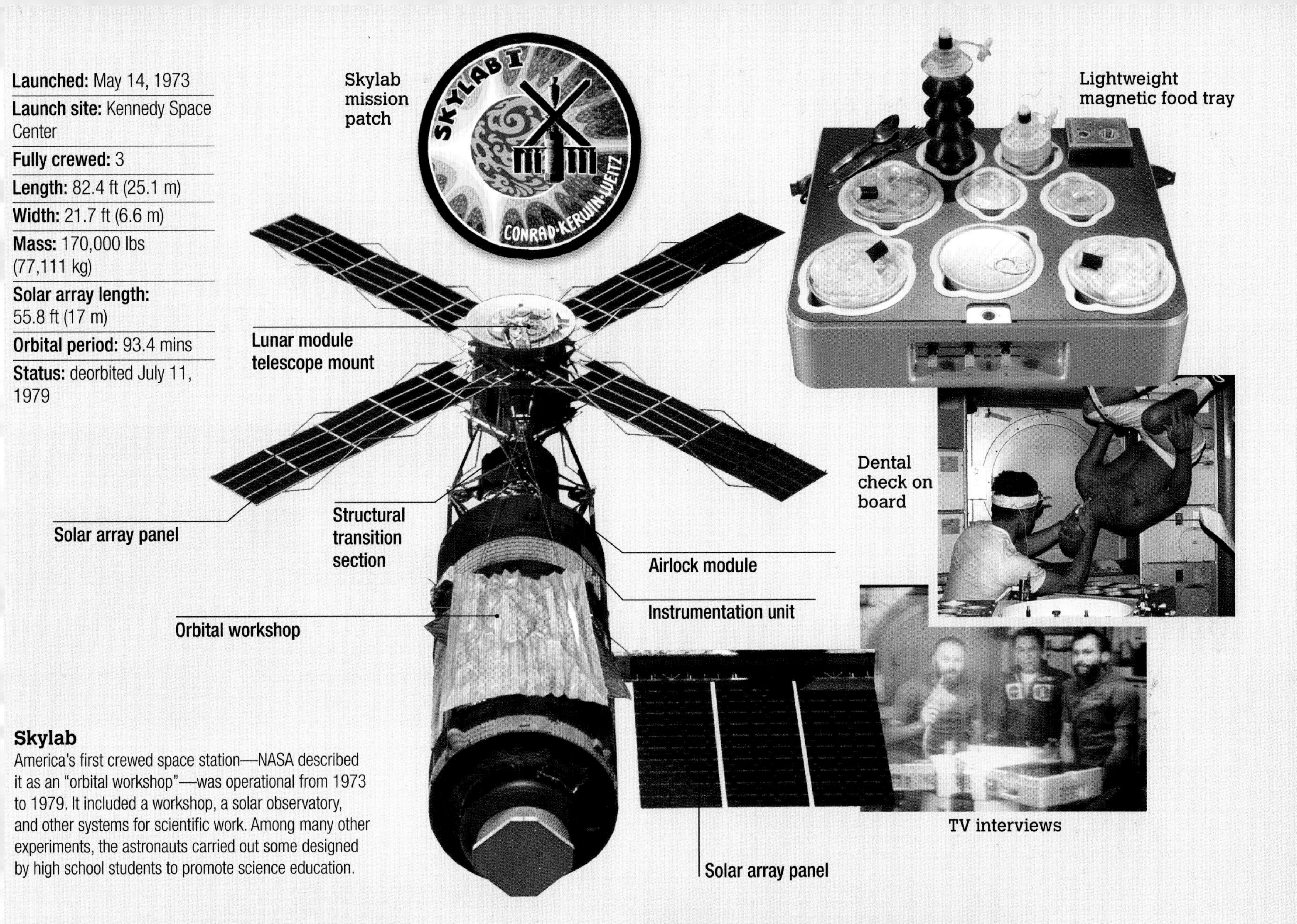

Launched: May 14, 1973
Launch site: Kennedy Space Center
Fully crewed: 3
Length: 82.4 ft (25.1 m)
Width: 21.7 ft (6.6 m)
Mass: 170,000 lbs (77,111 kg)
Solar array length: 55.8 ft (17 m)
Orbital period: 93.4 mins
Status: deorbited July 11, 1979

Skylab

America's first crewed space station—NASA described it as an "orbital workshop"—was operational from 1973 to 1979. It included a workshop, a solar observatory, and other systems for scientific work. Among many other experiments, the astronauts carried out some designed by high school students to promote science education.

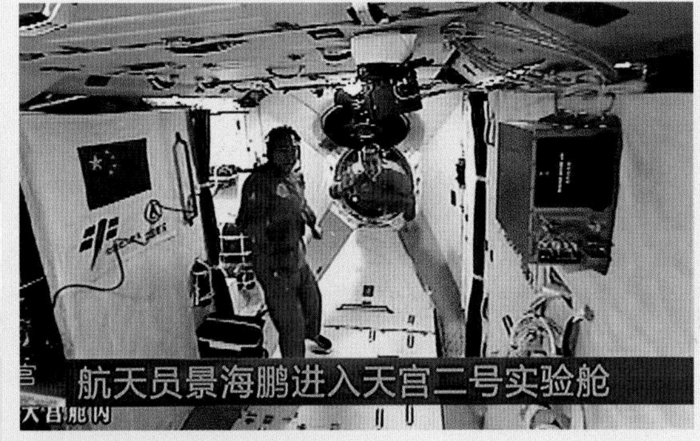

Launched: September 15, 2016
Launch site: Jiuquan Satellite Launch Center
Fully crewed: 2
Length: 34 ft (10.4 m)
Width: 11 ft (3.35 m)
Mass: 18,960 lbs (8,600 kg)
Solar array length: 55.5 ft (17 m)
Orbital period: 92 mins
Status: Operational

Tiangong 2

This Chinese space laboratory is a testbed for the technologies to be used when China builds a large modular space station sometime around 2022. On board, the astronauts conduct experiments from the cultivation of plants such as rice and cress, to scanning themselves to establish their bodies' fitness.

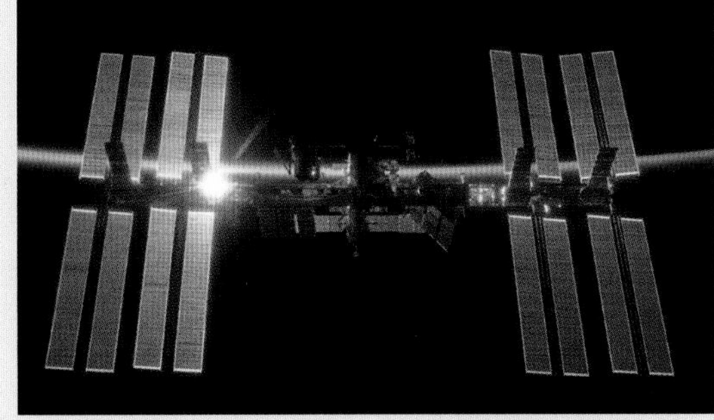

International Space Station (ISS)

The ISS is nearly four times as large as Mir and about five times as large as Skylab. The wingspan of its solar array is longer than the wingspan of a Boeing 777. It is made up of 16 modules—5 Russian, 8 US, 2 Japanese, and 1 European. In 2017, astronaut Peggy Whitson broke NASA's record for time in space on it—665 days, a world record for Americans.

Built: 1998–2011
Launch sites: Baikonur Cosmodrome and Kennedy Space Center
Fully crewed: 6
Length: 240 ft (73 m)
Width: 356 ft (108.5 m)
Mass: 925,355 lbs (419,725 kg)
Solar array length: 239.4 ft (73 m)
Orbital period: 92.65 mins
Status: Operational

SPACE SHUTTLES

SPACE TECHNOLOGY

For 135 missions, astronauts strapped themselves into a shuttle knowing that it was able to carry them to space and back—the first spaceship to do this. Every part of the ship could be used again, apart from the external fuel tank that burnt up in the atmosphere. The shuttle fleet clocked up more than 513 million miles (826 million km) ferrying astronauts and cargo, often destined for the ISS.

Lift off!

This remarkable craft took off like a rocket, reaching 17,500 mph (28,200 km/h) inside eight minutes. To lift it into orbit, it was attached to two solid booster rockets, had three main engines, an external fuel tank, and an orbital maneuvering system (OMS). The boosters provided most of the thrust and were jettisoned and parachuted back down to Earth for reuse.

NASA SHUTTLE FLIGHTS

ENTERPRISE test shuttle
No. of flights: 5 test flights, not in space
First flight: August 12, 1977
Last flight: October 26, 1977

COLUMBIA
No. of flights: 28
First flight: April 12, 1981
Last flight: February 1, 2003

CHALLENGER
No. of flights: 10
First flight: April 4, 1983
Last flight: January 28, 1986

DISCOVERY
No. of flights: 39
First flight: August 30, 1984
Last flight: February 24, 2011

ATLANTIS
No. of flights: 33
First flight: October 3, 1985
Last flight: July 8, 2011

ENDEAVOR
No. of flights: 25
First flight: May 7, 1992
Last flight: May 16, 2011

All-round view
These windows gave a view of Earth, while rear windows looked out on the cargo bay.

Crew
The commander sat on the left and the pilot on the right.

Flying the shuttle

The commander and pilot flew the shuttle like airplane pilots. Controls behind their seats on the flight deck were used for maneuvering for rendevous with the ISS, or operating remote manipulator systems, for example to deploy satellites.

Underside of shuttle wing showing debris damage

Detail of damage to thermal tile

Running a risk

Space debris caused regular wear and tear for the shuttles. Then, in 2003, NASA suspended flights for more than two years because of the loss of Columbia and its crew. They investigated this and an earlier fatal accident of the shuttle Challenger.

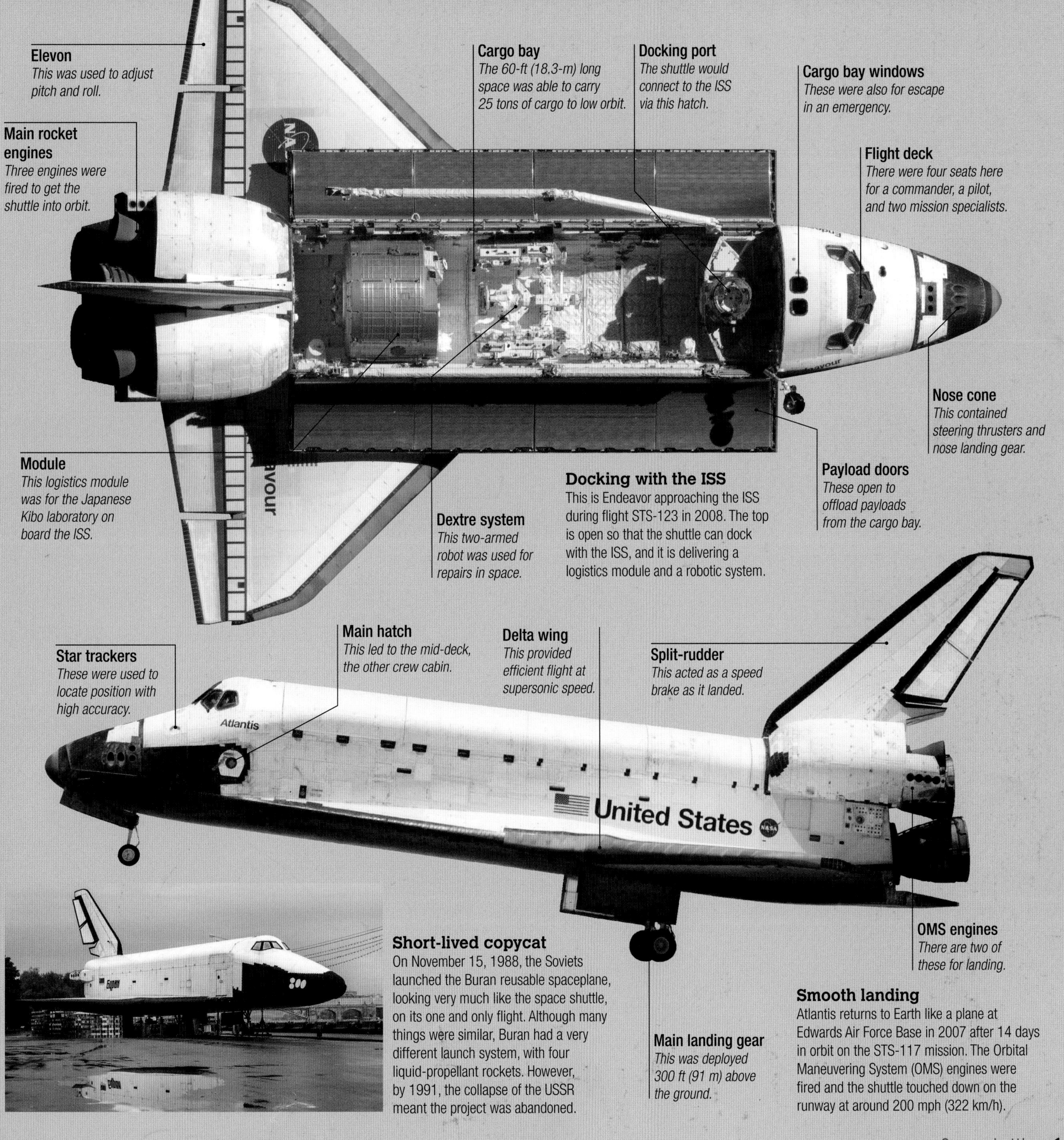

Docking with the ISS
This is Endeavor approaching the ISS during flight STS-123 in 2008. The top is open so that the shuttle can dock with the ISS, and it is delivering a logistics module and a robotic system.

Short-lived copycat
On November 15, 1988, the Soviets launched the Buran reusable spaceplane, looking very much like the space shuttle, on its one and only flight. Although many things were similar, Buran had a very different launch system, with four liquid-propellant rockets. However, by 1991, the collapse of the USSR meant the project was abandoned.

Smooth landing
Atlantis returns to Earth like a plane at Edwards Air Force Base in 2007 after 14 days in orbit on the STS-117 mission. The Orbital Maneuvering System (OMS) engines were fired and the shuttle touched down on the runway at around 200 mph (322 km/h).

Atlantis
Canada™

Atlantis in space

This is the space shuttle Atlantis docking with the ISS during mission STS-132 in 2010. On the right is ESA's space laboratory Columbus. Atlantis had launched satellites, the Magellan and Galileo probes, and the Compton Gamma-Ray Observatory, as well as flying missions to the space stations Mir and the ISS. The final flight of the whole space shuttle program was carried out by Atlantis, which launched on July 8, 2011, on the STS-135 mission. It touched down safely on July 21, 2011 and the space shuttle era was brought to a close.

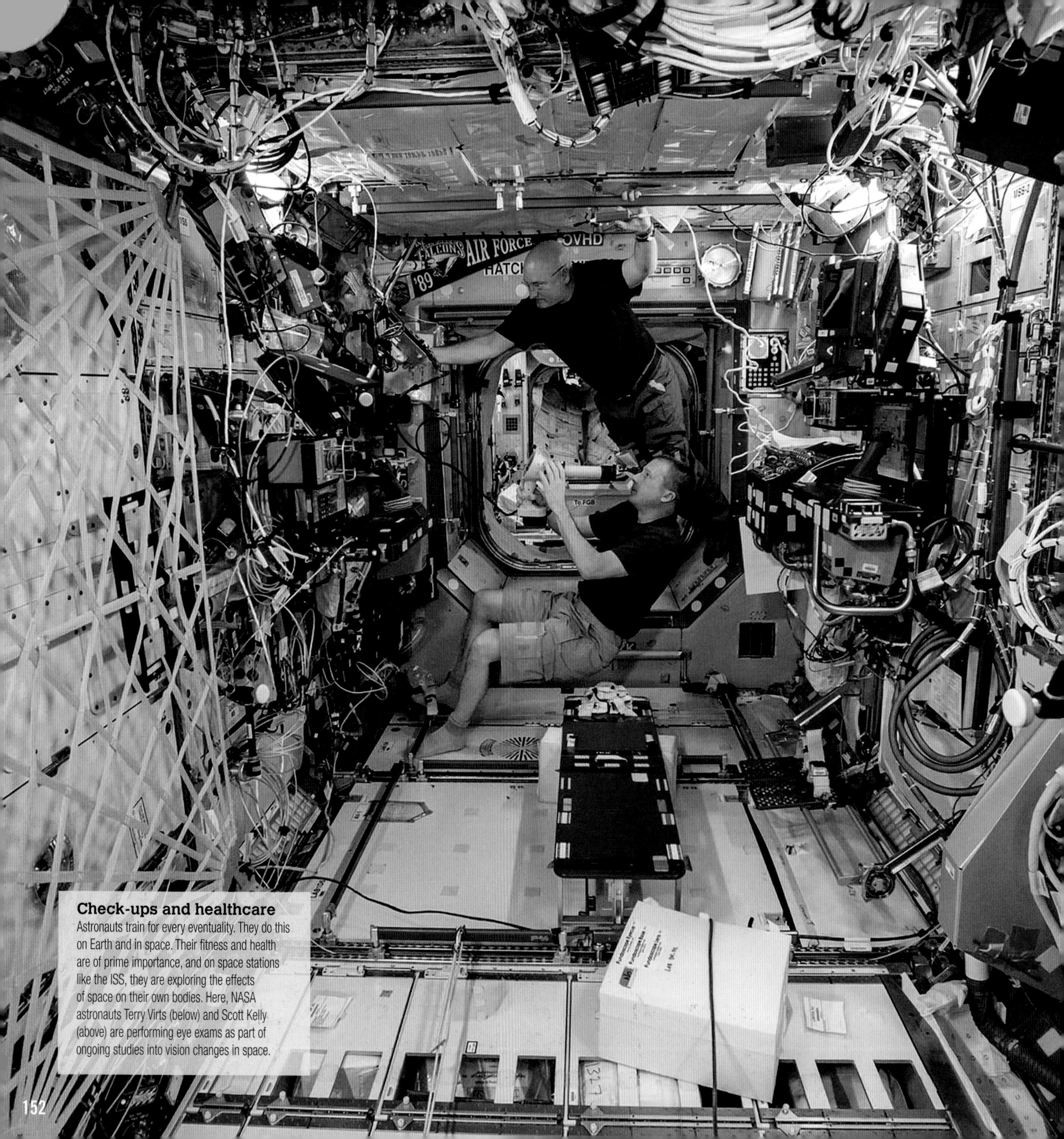

Check-ups and healthcare
Astronauts train for every eventuality. They do this on Earth and in space. Their fitness and health are of prime importance, and on space stations like the ISS, they are exploring the effects of space on their own bodies. Here, NASA astronauts Terry Virts (below) and Scott Kelly (above) are performing eye exams as part of ongoing studies into vision changes in space.

RIDE INTO SPACE

BY JERRY L. ROSS, SHUTTLE ASTRONAUT AND MISSION SPECIALIST TO MIR AND THE ISS

You never forget a ride into space—it is so exciting. But it would be crazy not to feel a little anxious as well. Launch and reentry are the times of greatest risk.

Three hours before takeoff, the crew are strapped into their seats on top of the rocket that contains 3.8 million lbs (1.7 million kg) of explosive fuel. The launch pad close-crew closes the hatch and performs the final checks. The crew know the physics: a rocket has to reach a very high speed in a few minutes to achieve orbital velocity of 17,500 mph (28,200 km/h). The main engines ignite six seconds before liftoff, and the ship starts to rattle and shake. When the solid rocket boosters ignite for launch, it feels like a kick in the back as the shuttle speeds away from the launch pad.

The rocket rumbles and shudders. At about 40 seconds after liftoff, the engines throttle down to reduce stress and then throttle back up to full thrust. At about two minutes after liftoff, the solid rocket boosters have each consumed 1.1 million lbs (499,400 kg) of propellant, and there is a bang and a flash of bright rocket firings as the empty boosters are pushed off the external tank.

During the last minute of the ascent into orbit, the crew is subjected to an acceleration three times the force of gravity. This feels like someone very large is sitting on your chest and makes it hard to breathe. But eight and a half minutes after launch the rocket engines stop, the crew members are floating in their seats, and they are in space.

Astronaut
Jerry L. Ross flew seven Space Shuttle missions and jointly holds the record for the most spaceflights. He launched satellites, spacewalked nine times, supplied Mir, and helped build the ISS.

Spacewalks are another time when everyone must be especially alert. In 2001, Chris Hadfield was working on the outside of the ISS when his left eye started to burn and water. In the absence of gravity, the tears did not flow down his cheek but became a blob that spread to his other eye, blinding him temporarily. This severe eye irritation was due to the anti-fog solution used on his spacesuit visor that affected his sight and forced him to vent oxygen into space.

Returning to Earth is just as exciting and almost as dangerous as the launch. Thrusters fire to slow down the ship, then it glides into the increasingly thick atmosphere. The movement creates a plasma flow over the ship—astronauts see a fireball as the temperature rises to 3,000°F (1,648°C) on the outside of the craft because of the shockwaves and friction. Then the flames die away and the ship flies down at a steep angle to land on the runway.

"I got to **operate** the **robotic arm**; do **experiments**, satellite launches, **spacewalks**; help build the **ISS**."

Piloting the shuttle

Keeping fit and doing health checks

Conducting experiments

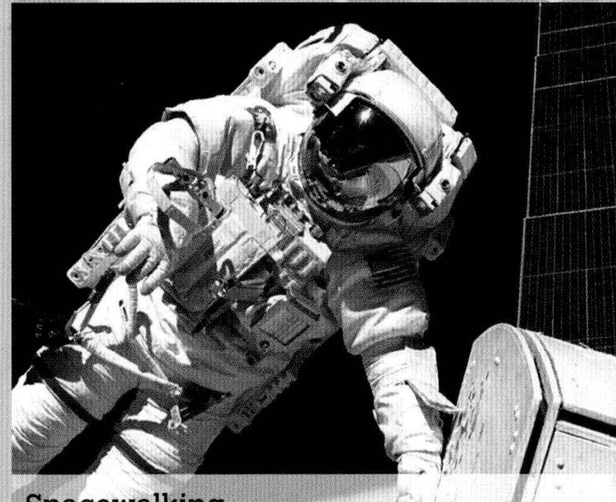

Spacewalking

ABOARD THE ISS

SPACE TECHNOLOGY

The weightlessness of microgravity makes everyday tasks tricky on board a space station. Astronauts cannot risk having droplets of soapy water or grains of salt floating around to clog up equipment or get stuck in someone's eye. Every work day is run to a tight schedule that might include running experiments, fixing a solar panel, and two hours of special exercises to stop bones and muscles weakening due to a lack of gravity. But there is also time to look out of a window and watch a sunrise or sunset while speeding around Earth.

The **ISS weighs** nearly **420 tons** and covers the **area** of a **soccer field.**

Anatomy of a space station

The ISS consists of a group of pressurized modules that are laboratories, utility hubs, living quarters, cargo spaces, or docking ports. Kibo, the Japanese laboratory, is the largest of the modules. Attached to the group of modules is a long, main truss that supports giant solar panels, heat radiators, external manipulating arms, and other equipment.

ISS in 2000, when the first crew arrived

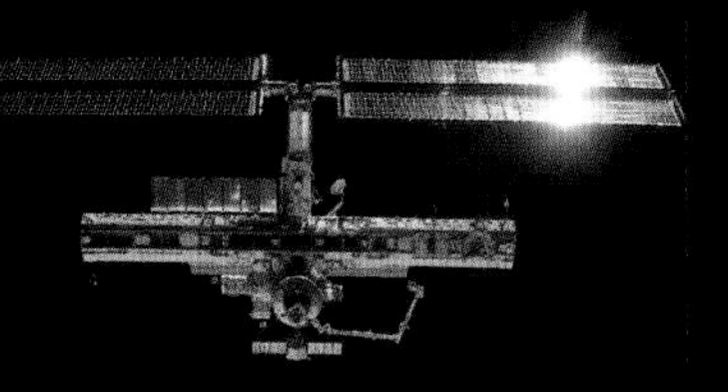

ISS in 2002 with first photovoltaic panel in place

ISS in 2007 with two more pairs of solar arrays and extra modules

Assembling the ISS

The ISS was built in orbit, piece by piece. The first component, Russia's Zarya—used for communications, power, and storage—was launched in 1998, and the structure has grown ever since. It received its first full-time crew of three astronauts in 2000. As of January 2018, 230 people from 18 countries have visited the ISS.

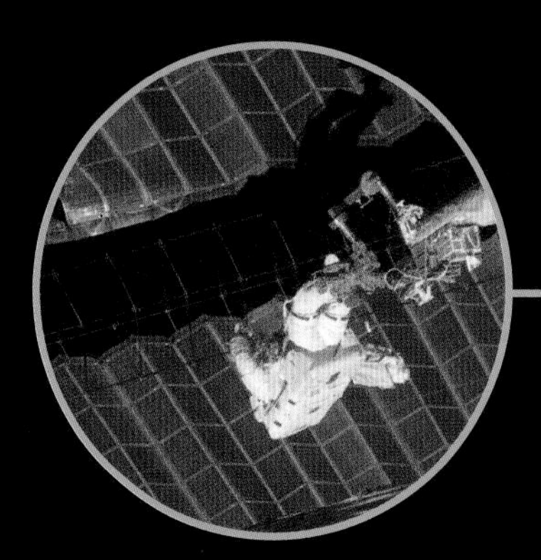

Panel repair

The ISS has to be maintained both inside and out. It has equipment that needs replacing or upgrading on a regular basis, and its fabric has to be checked and repaired when it is damaged or simply worn out. Astronauts make scheduled and, where necessary, unscheduled spacewalks to work on the outside of the space station.

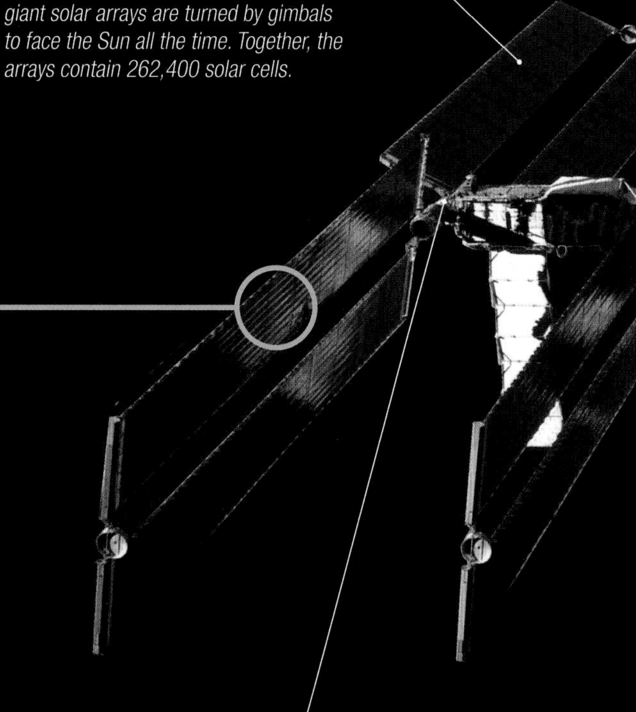

Starboard photovoltaic array
Sunlight is plentiful in space, and these giant solar arrays are turned by gimbals to face the Sun all the time. Together, the arrays contain 262,400 solar cells.

Truss segment
The ten trusses are connected to one another and form the main structure of the ISS. They support the solar panels and other equipment.

Alpha magnetic spectrometer (AMS)
This physics detector measures and analyzes cosmic rays. It has identified nine million cosmic ray events as electrons or positrons (antimatter).

Docked on

There are docking ports on the different modules of the ISS, so several spacecraft can dock at any one time. The shuttles (right) docked here on the NASA port when they delivered astronauts and cargo to the ISS.

Life on board

The working day on the ISS might include running an experiment, talking to some schoolkids, or walking in space. Astronauts must also exercise on average two hours a day. They can then chat with their families and play a few games before settling down in a sleeping bag inside the one-person cabins.

Experiments as part of the regular routine

Magnetic food tray

Eating food on board can prove tricky

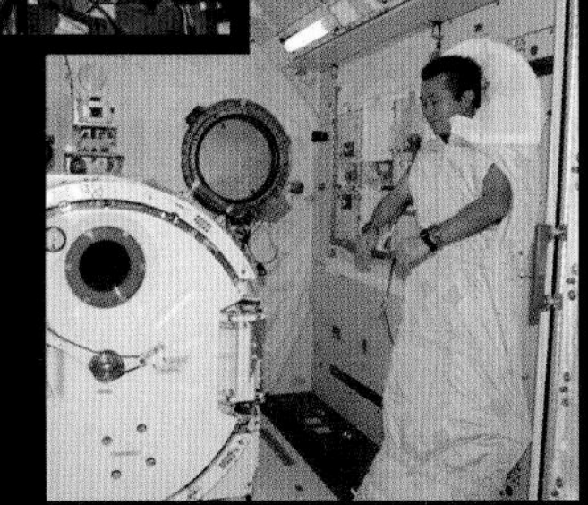

Beds fixed to the wall to avoid floating away

Harmony node
This acts as an internal connecting port to the international science labs and provides air and other systems.

Columbus laboratory
The first European research facility in space, ESA's science laboratory was attached in 2008.

Kibo research module
The Japanese Experiment Module (JEM) is Japan's first human space facility.

Destiny laboratory
This was installed in 2001 and is used for US experiments in human life sciences, physical sciences, and Earth observation.

Heat rejection radiators
Each radiator has seven panels that extend to 50 ft (15 m) in length. Ammonia is passed through a series of tubes to collect excess heat from electronic systems and the radiators dissipate it into space.

Zvezda crew quarters
This was one of the early modules in space and provided the living quarters for all the astronauts at first.

Quest airlock store
This stores spacesuits and equipment. The astronauts stay here in a reduced nitrogen atmosphere before a spacewalk. Quest is one of two segments. The next compartment is the "Crew lock" from which astronauts can exit into space and then return to after they have done their work outside.

Pump module for thermal control radiators
This pumps ammonia through giant radiators, transporting excess heat away and cooling equipment around most of the modules.

REACHING

THE STARS

FINDING OUR PLACE

The two Voyager probes that have taught us so much about our solar system are now at its edge and, over the coming years, will continue to send back data about their road trip through unexplored territory. In the future, we may develop technologies that will allow people to leave our solar system to fully explore the Milky Way and other galaxies, or even find other worlds on which they can live.

COSMIC PUZZLES

Although we cannot yet visit other galaxies, we can see them spectacularly clearly through ever-improving telescopes. The James Webb Space Telescope, to be launched in 2020, has a mirror that is seven times more powerful than that of Hubble (*see pp.40–41*), and astronomers think that it will help us see even more, including the stars and galaxies of the early Universe and many other wonders of star birth and death in deepest space. And scientists have developed some mind-blowing ideas—for example, that the Universe is expanding ever faster (and no spaceship will ever catch up) and that there are black holes that even light cannot leave—and these will be tested over time by improved technology.

AN ENDLESS SEARCH

Nearer home, our Sun is on a journey of its own that will one day end in a spectacular expansion that will wipe our planet off the galactic map. This may explain why humans feel driven to try to communicate with lifeforms on other worlds, or find places that the human race can use as alternative homes to Earth if needed. As science fiction writer Arthur C. Clarke wrote, "Sometimes I think we're alone in the Universe and sometimes I think we're not. In either case the idea is quite staggering."

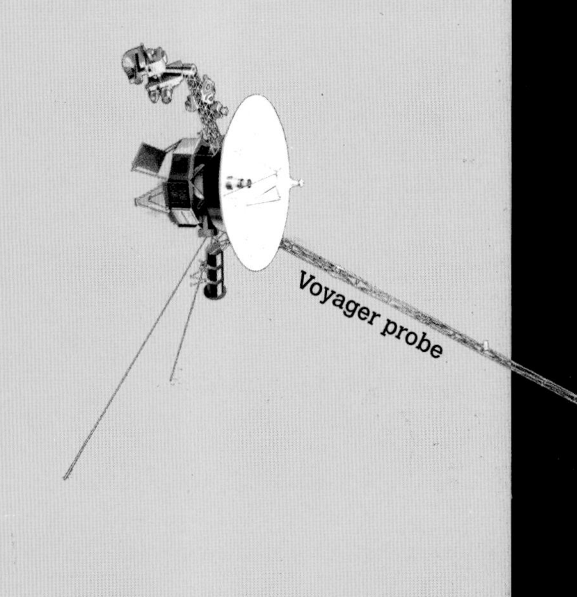

VOYAGERS' JOURNEYS

SPACE TECHNOLOGY

The twin Voyager 1 and 2 spacecraft are farther out in space than anything from Earth has ever been. After major discoveries about our solar system—from volcanoes on Jupiter's moon Io to the makeup of Saturn's rings—Voyager 1 has reached interstellar space. Vintage electronics and thrusters keep these van-sized probes zooming along and sending back data.

Plasma wave instrument
Used to measure the density, pressure, and speed of plasma.

Ultraviolet spectrometer
This scans for ultraviolet waves from near and distant objects in space.

Cosmic ray instrument
This measures energy in the probe's immediate environment.

Calibration target and radiator
This calibrates (standardizes) images and controls heat levels.

High-gain antenna
The 12 ft (3.7 m) wide reflector dish receives commands from Earth, and its radio signals are used to track the probe's position.

Radioisotope thermoelectric generator
This supplies the probe with electric current, provided by the decay of radioactive material.

Magnetometer boom
The magnetometers on this 42.5 ft (13 m) fiberglass boom sense magnetic fields in the immediate envionment.

Planetary radio astronomy and plasma wave antenna
The antenna senses radio waves from planetary systems and measures plasma waves in the immediate environment.

Uranus' orbit
Neptune's orbit
Voyager 1
Earth's orbit
Jupiter's orbit
Voyager 2
Saturn's orbit

Perfect timing

It was crucial in the planning that the geometric arrangement of the solar system's outer planets orbits was such that it allowed the Voyagers to swing from one planet to the other without needing large onboard propulsion systems.

Film star

A Voyager probe plays a starring role in the 1979 *Star Trek: The Motion Picture*, when a damaged Voyager 6 is found by an alien race.

Voyager 1
Voyager 2

Bow shock
A shockwave created by the movement of the heliosphere.

Heliosheath

Heliopause

Termination shock
A boundary marking the limit of the Sun's influence.

On the edge

Voyager 2 is currently in the heliosheath. This is the place where the solar wind is slowed by the pressure of interstellar gas. Its boundaries are the termination shock (innermost) and the heliopause (outermost). Voyager 1 has crossed the termination shock and is in interstellar space.

One of two

The two probes are identically designed and originally built to last five years. Launched in 1977, they are still going strong! Each probe has only 64kb of memory. Today's smartphones have 200,000 times more than that!

Recorded info

Fitted to the outside of both Voyagers are golden disks, two records with sounds and images selected to reflect life and culture on Earth.

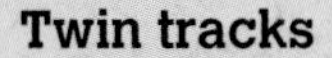

Twin tracks

The two Voyagers had the same primary mission—the exploration of Jupiter and Saturn. Their mission was then extended and their paths diverged. Voyager 2 went on to explore Uranus and Neptune. Both Voyagers are now exploring the edge of the solar system.

Aug. 20, 1977
Voyager 2 launches

Sept. 5, 1977
Voyager 1 launches

March 5, 1979
Voyager 1 approaches Jupiter

July 9, 1979
Voyager 2 approaches Jupiter

Nov. 9, 1980
Voyager 1 nears Saturn and its moon Titan

Aug. 25, 1981
Voyager 2 nears Saturn and some of its moons

Jan. 24, 1986
Voyager 2's closest approach to Uranus

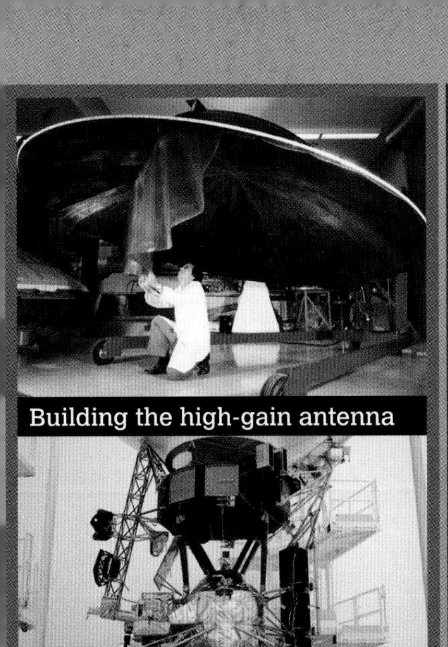

Building the high-gain antenna

Voyager test model

Assembling Voyager 2

Launch of Voyager 1 on Titan III

Building a project

The probes were built at the Jet Propulsion Laboratory in California. They were originally called Mariner 11 and 12, but then moved to their own program and were renamed.

Recording session

Gold-plating process

One of the pictures on the disk

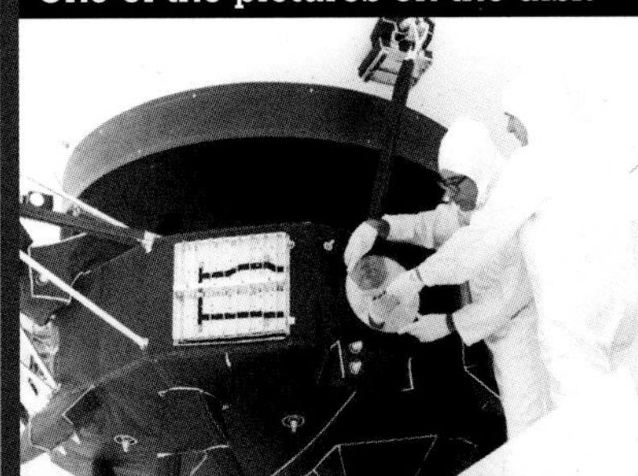

Attaching the disk to Voyager

Golden disk manufacture

The gold-plated records were each encased in a protective aluminum jacket. Symbols are used to explain where the probes have come from and how the record is to be played.

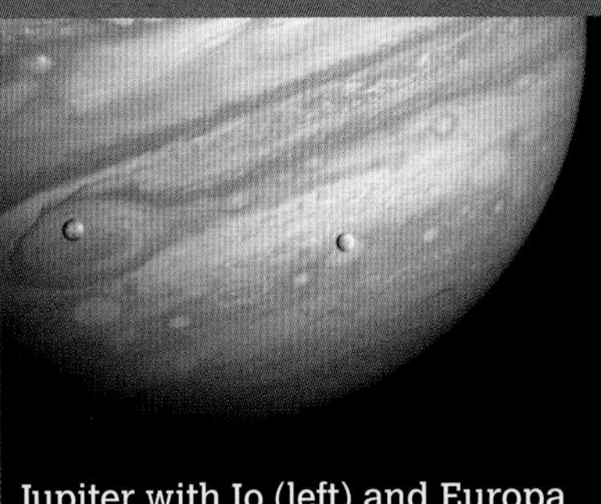

Jupiter with Io (left) and Europa

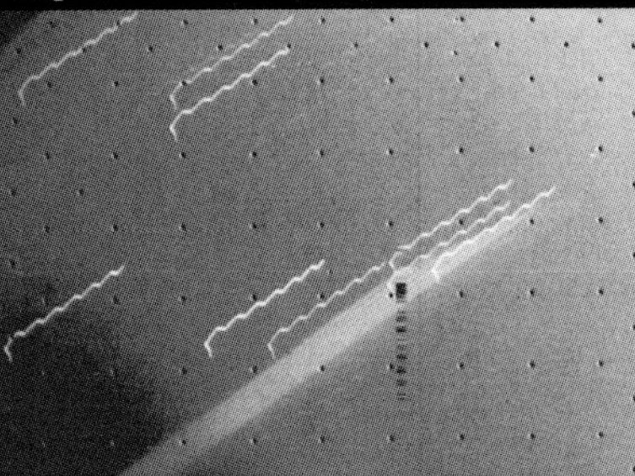

First evidence of Jupiter rings

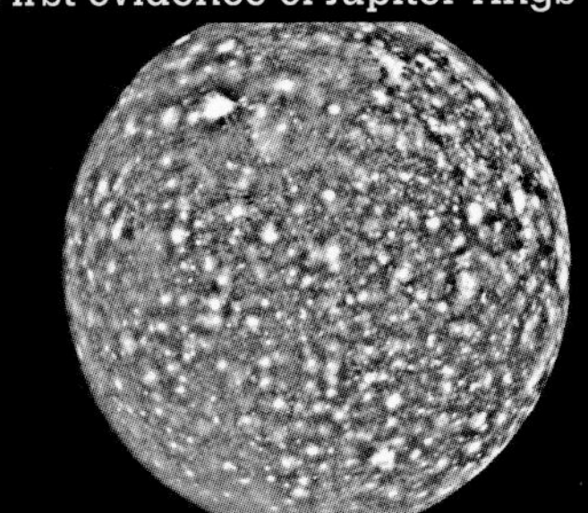

Jupiter's moon Callisto

"Portrait" of the solar system

Voyager 1 highlights

This probe found active volcanoes on Io and sent back the first hi-res images of Saturn's rings. As it left the solar system, it also looked back for a "portrait" of six of the planets.

Jupiter's moon Ganymede

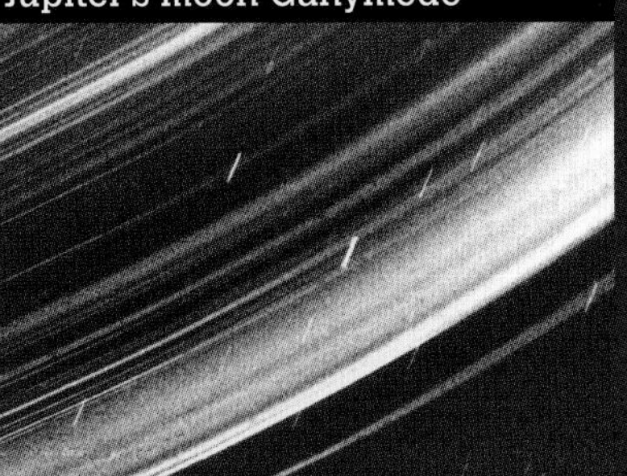

Rings of Uranus

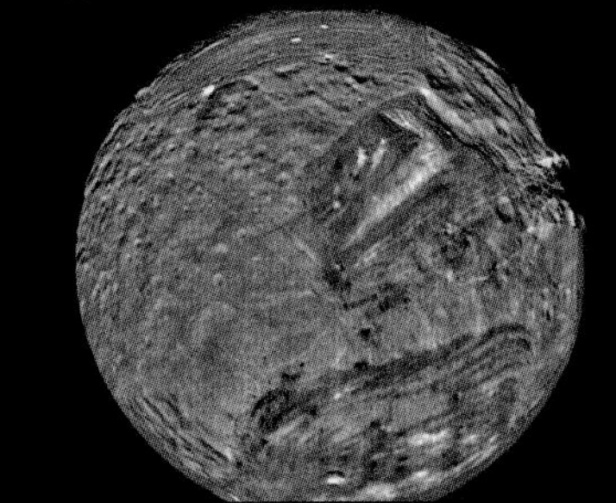

Uranus' moon Miranda

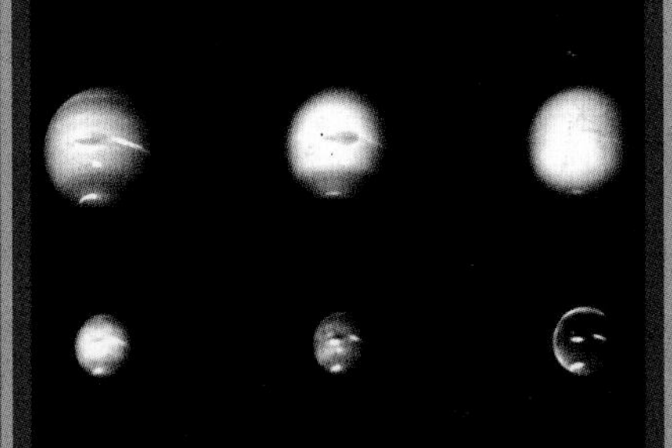

Views of Neptune

Voyager 2 highlights

Alongside the many observations of the moons of Jupiter and Saturn, this probe has sent back the only closeup views of Uranus and Neptune that humans have ever seen.

Aug. 25, 1989
Voyager 2's closest approach to Neptune

Oct. 10 and Dec. 5, 1989
Voyager 2's cameras turned off to save power

February 14, 1990
Voyager 1's cameras turned off to save power

Feb. 17, 1998
Voyager 1 becomes farthest human-made object from Earth

Dec. 16, 2004
Voyager 1 crosses into the heliosheath

Aug. 30, 2007
Voyager 2 crosses into the heliosheath

August 25, 2012
Voyager 1 enters interstellar space

GALAXY TYPES

GALAXIES

The Voyager spacecraft are the first human-made objects to head outside our solar system across our galaxy, the Milky Way. When we marvel at our solar system's vast distances, it is easy to forget that our Sun is only one of up to 400 billion stars in the Milky Way, itself only a medium-sized galaxy. There are probably billions or even trillions of galaxies in the Universe.

Invisible galaxies
Hubble's Ultra Deep Field revealed in 2016 that there may be ten times as many galaxies as the 200 billion that astronomers previously thought. Ninety percent of the galaxies in the Universe have yet to be studied.

Speeding galaxy
NGC 1472A is a dwarf irregular galaxy around 71 million light-years away from Earth in the constellation Eridanus. The galaxy is speeding toward the center of the Fornax cluster, a group of 58 galaxies. The gravities of the other galaxies have pulled it into this shape.

IRREGULAR

CHAOTIC GALAXIES

These do not have a distinct shape. Some have always been like this, others changed as they passed through another galaxy. Only about five percent of galaxies fall into this category. They are made up of a mix of old and new stars, and are often fairly small.

Rare blue galaxy
This is a blue compact dwarf galaxy, NGC 5253, about 12 million light years from Earth in the constellation Centaurus. These galaxies contain molecular clouds without dust or heavier elements. The clouds are similar to those that formed the first stars in the early Universe.

Star birthplace
UGC 4459 is rich with young blue stars and older red ones and is made up of several billion stars. This small galaxy is about 11 million light-years away from Earth, in the constellation Ursa Major.

ELLIPTICAL

STRETCHED-OUT CIRCLES

The largest galaxies are elliptical. Most are oval in shape, but they are sometimes almost circular, and occasionally stretched out like a tube. They usually have many stars clustered together near the center, and they hold little gas or dust and so rarely produce new stars.

Bright light in the sky
This massive galaxy, M87, can be seen from Earth. Its jet (right of center) extends out into space 5,000 light-years. It is 53 million light-years away from Earth and is in the constellation Virgo.

Galactic merger
NGC 3597 is the result of a collision between two large galaxies, and it is slowly becoming a giant elliptical galaxy. It is around 150 million light-years away from Earth in the constellation Crater.

Some **elliptical galaxies** contain **supermassive black holes** seven billion times **heavier** than the **Sun**.

SPIRAL

SWIRLING STARS

Around a fifth of galaxies, including our own Milky Way, are flat, rotating disks with a bulge in the middle and at least two long spiral arms swirling outward. Some of these are called "barred" spirals because the bulge looks elongated, like a bar. At the center of spiral galaxies are older stars. The rotating movement of the spiral arms creates waves in which new stars often form.

Glittering disk
Viewed almost exactly side-on, the flat, full length of the galaxy can be seen, with the stronger glow of the central bulge. This is ESO 121-6, which lies in the constellation Pictor.

On its own
Most galaxies are in groups, but this one, NGC 6503, is on its own at the edge of the Local Void. This area of space is some 18 million light-years away from Earth in the constellation Draco. The bright blue regions contain newly forming stars.

Pinwheel galaxy
Messier 83 is the 83rd entry in the catalog of bright objects that the comet-hunting astronomer Charles Messier made in the 1700s. It is also known as the Southern Pinwheel Galaxy and is around 15 million light-years away from Earth in the constellation Hydra.

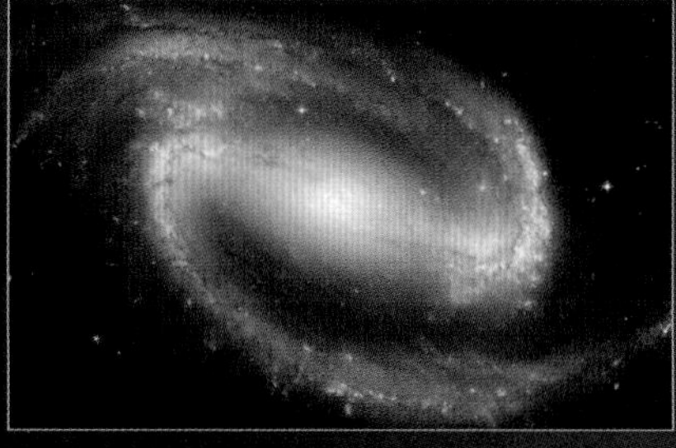

Barred spiral galaxy
This large barred spiral, NGC 1300, is around 70 million light years away on the edge of the constellation Eridanus. The nucleus itself is some 3,000 light-years across and thought to have a supermassive black hole in its center.

Types of spiral
These are the variations in the shape of spirals (Sa, Sb, Sc) and barred spirals (SBa, SBb, SBc) and how their arms are arranged.

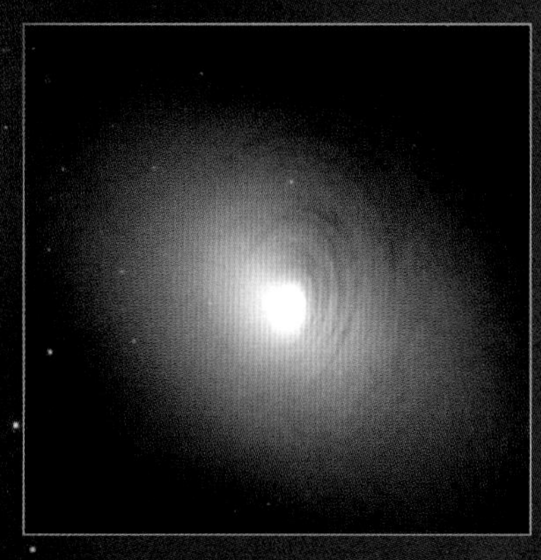

Variation on a theme
Lenticular, or SO, galaxies have a bulge and a thin disk, but they do not have a spiral structure. This one is the barred lenticular galaxy NGC 2787, which is around 25 million light-years away in the constellation Ursa Major.

THE MILKY WAY

GALAXIES

Our galaxy, the Milky Way, is organized into spiral arms of giant stars that light up interstellar gas and dust. It takes 230 million years to circle its own center. Scientists have long known there is a supermassive black hole at its center—Sagittarius A, 400 times bigger than the Sun. A recent study suggests that it may be host to thousands of black holes.

Central bulge resembles fried eggs

Egg-like galaxy
The part of the Milky Way you can see from Earth on a clear night depends on which hemisphere you are in. It (and other spiral galaxies) looks very like two fried eggs laid back to back. The central bulge is the yolks, and the rotating disk with its spiral arms is the whites.

Pleiades
This bright star cluster can be seen from nearly everywhere on Earth, even the North Pole.

NGC 7789
An open cluster that is also called "Caroline's Rose" for Caroline Herschel.

Andromeda galaxy
This is the Milky Way's nearest major galaxy *(see p.169).*

The Sun
Our Sun lies some 27,000 light-years away from the central bulge of the Milky Way, on a branch of the Sagittarius arm.

Alpha Cygni
Also known as Deneb, or the Pole Star, this is a bright blue supergiant.

Vega
A blue-white star, Vega is close to the Sun at a distance of only 25 light-years.

The Great Rift
These dark dust clouds form a long river that splits the bright band of the Milky Way.

NGC 6822
This barred irregular galaxy, also known as the Bubble Galaxy, is near the Milky Way.

Bright ribbon in the sky
This stunning image of the Milky Way as it is seen from Earth is the work of the European South Observatory (ESO) in the cold, clear atmosphere of the highlands of Chile. This 360-degree panorama reveals the galaxy that we are part of in all its glory. We see the Milky Way edge-on, with the disk of the galactic center glowing brightly in the middle.

To the **Maori** the **Milky Way** is the **wake** of the **canoe** of the **mythical warrior Tama-rereti.**

The galaxy

The Milky Way is a barred spiral galaxy (*see p.163*). It has four main arms: Norma, Sagittarius, Perseus, and Scutum-Centaurus, and it is there that most of the stars are born (*see p.170*)—Perseus and Scutum-Centaurus in particular seem to contain an abundance. Our Sun is on a minor arm called the Orion spur. The diameter of the disk is 100,000 light-years and the bar at the center is around 27,000 light-years across.

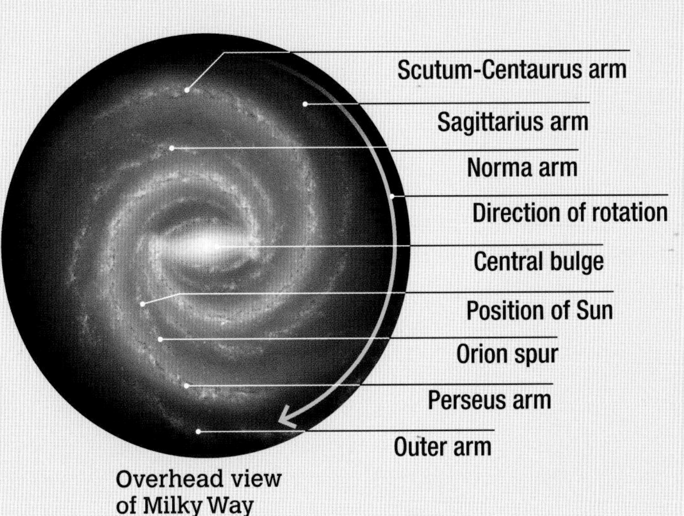

Overhead view of Milky Way

Las Campanas Observatory, Chile

Apache Point Observatory, New Mexico

Observing the Milky Way

Las Campanas houses the Giant Magellan Telescope, which has let astronomers study stars across the whole Milky Way in more detail than ever before. At Apache Point, the APOGEE-2 project is finding out about the early Milky Way by examining the stars in the central bulge.

Central bulge
The galactic center of the galaxy is a round structure made up of old stars, gas, and dust.

Small Megallanic cloud
This dwarf irregular galaxy can be seen from the southern hemisphere.

Alpha centauri
This is the closest star system to our solar system and contains a planet orbiting the star Proxima Centauri.

NGC 3201
This star cluster may have a black hole four times the mass of our Sun.

Large Megallanic cloud
This irregular barred galaxy is the third closest galaxy to the Milky Way.

Canopus
Also known as Alpha Carinae, this is the second-brightest star in the night sky.

M79
This globular cluster contains some of the oldest stars in our galaxy.

Orion nebula
Orion has two of the very brightest stars seen from Earth, Betelgeuse and Rigel.

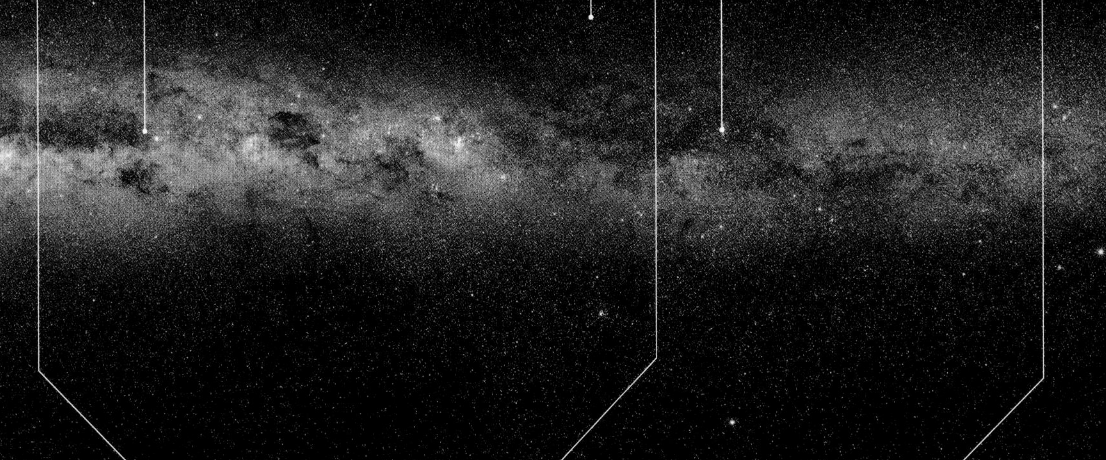

Galaxy type: barred spiral
Constellation: Sagittarius
Group: Local Group
Age: 13.6 billion years
Diameter: 100,000 light-years
Number of stars: estimated up to 400 billion

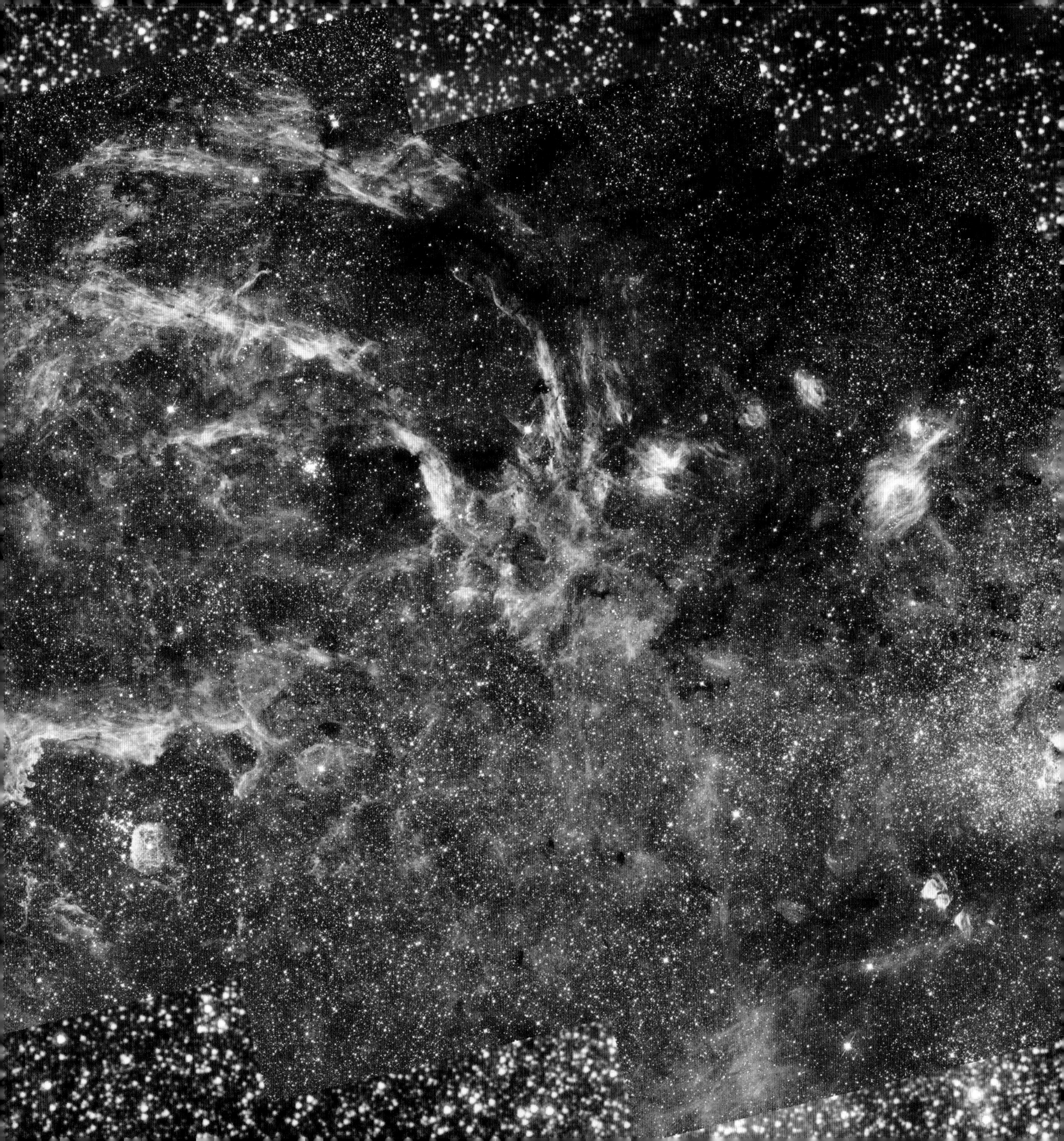

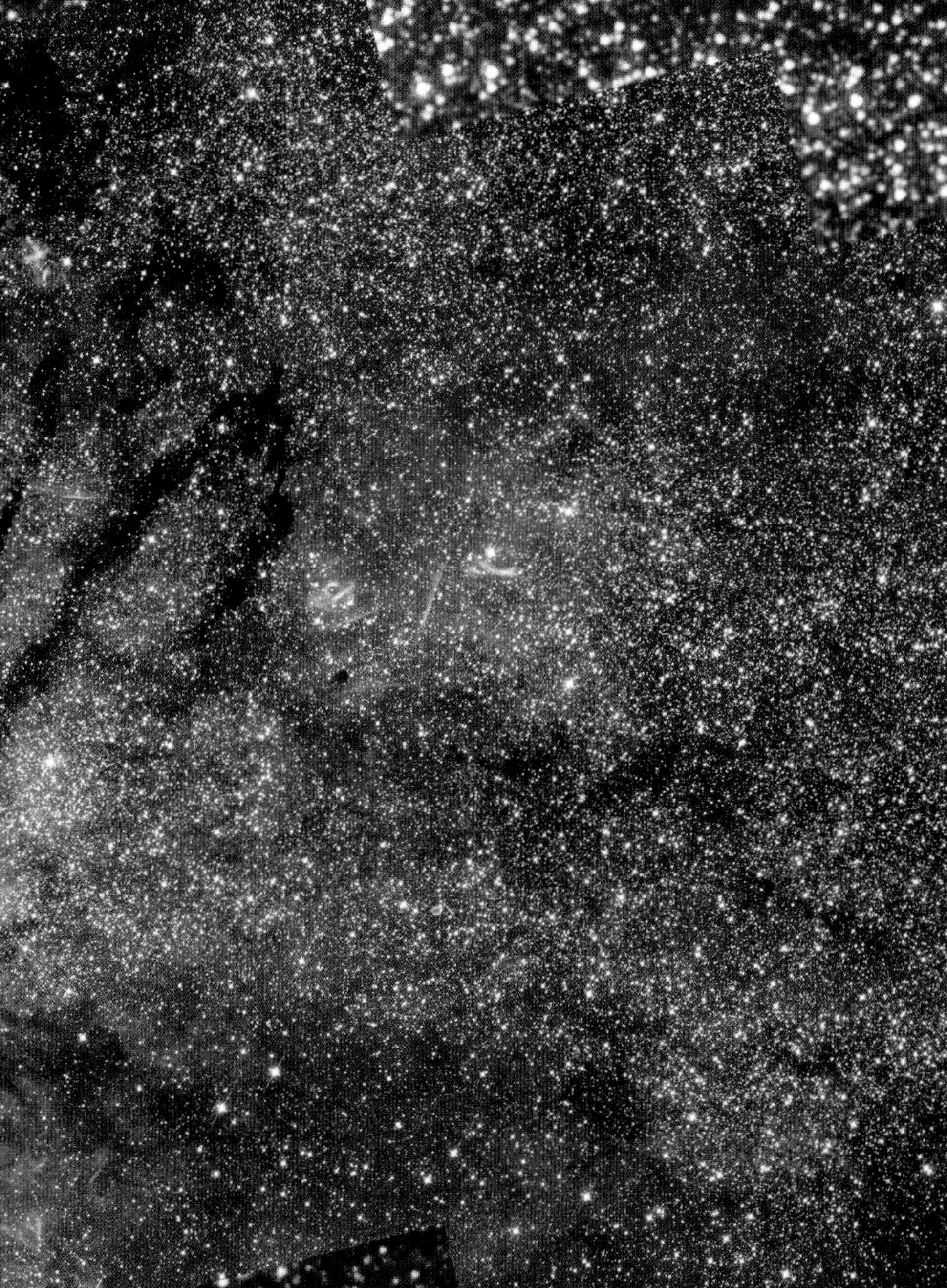

Visible—the center of our galaxy

This extraordinary view of the center or core of the Milky Way galaxy is a mosaic made up of 2,304 images achieved by 144 Hubble orbits. Hubble's Near Infrared Camera and Multi-Object Spectrometer (NICMOS) images were combined with color images from a previous Spitzer Space Telescope survey using its Infrared Astronomy Camera (IRAC). The central 300 light-years—the white core—is surrounded by swirling hot, ionized gases. In visible light, the core would be obscured by dust clouds, but infrared light is able to penetrate.

NEAR NEIGHBORS

GALAXIES

Most galaxies cluster into groups, and our Milky Way is one of more than 30 that make up the Local Group. It is not a friendly arrangement—our galaxy is sucking in parts of the nearby Magellanic clouds, which are far smaller and therefore less powerful. Meanwhile, the Andromeda galaxy is hurtling towards us, ready for the lead role in the next chapter of star wars . . .

The Local Group
This contains mostly dwarf galaxies. The map charts the positions of the galaxies on these pages, and some of the galaxies nearest to them.The Milky Way and Andromeda alone make up more than half of the mass of the group.

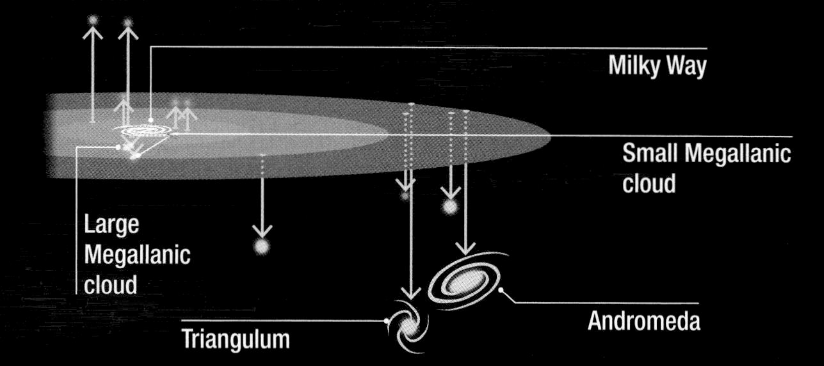

TRIANGULUM

SPIRAL GALAXY

This galaxy does not seem to have a supermassive black hole in the middle of its spiral. It is a star factory with vast ionized hydrogen clouds where new stars are born. The newcomers are very hot at 72,000°F (40,000°C).

Pinwheel in the sky
Nearly 3 million light years from Earth, Triangulum is the third largest galaxy in the Local Group. It is the most distant deep sky object visible to the naked eye. Its largest new stars are 120 times the mass of our Sun.

Also called: M33 or NGC 598
Type: spiral
Constellation: Triangulum
Group: Local Group
Diameter: 60,000 light years
Number of stars: 40 billion

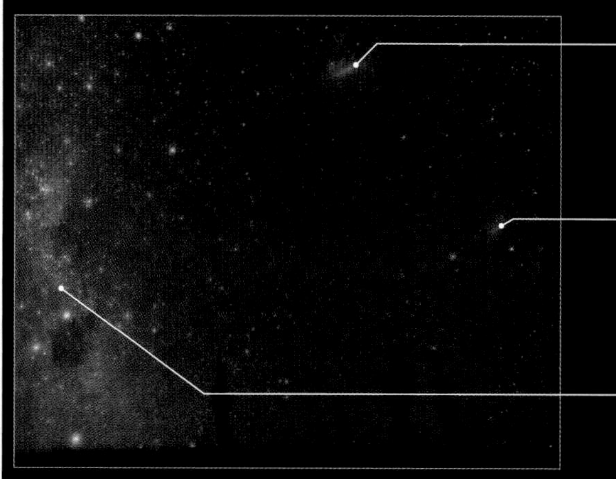

Large Magellanic cloud
This barred dwarf galaxy is 163,000 light years away.

Small Magellanic cloud
This dwarf galaxy is around 200,000 light years away.

Milky Way
Its gravity is pulling in gases from both clouds and disrupting their shapes.

Faint clouds
The two Magellanic clouds are visible to the naked eye in the southern hemisphere. They both have star-forming regions and may eventually collide with the Milky Way.

MAGELLANIC CLOUDS

IRREGULAR DWARF GALAXIES

The earlier spiral shapes of these dwarf galaxies—so-called because they are only 14,000 and 7,000 light years across—have been pulled apart by our galaxy's stronger gravity. They hold more hydrogen and helium than the Milky Way, but fewer metal elements. The larger one is home to a star nursery, the Tarantula Nebula.

At the core
In the center of the galaxy, a bright ring of older, cooler red stars surround a disk of young, hot, blue stars. These in turn swirl around a supermassive black hole.

Closest neighbor
The beautiful swirl of the blue-white rings of the Andromeda galaxy is where new stars are born. The image of the Milky Way's nearest neighbor was taken by NASA's GALEX space telescope.

Binary features
This false-color X-ray shows what are probably X-ray binary stars in the central area of Andromeda. Binary stars rotate around each other—one is a normal star, while the other is either a neutron star or a black hole.

Halo detail of distorted galaxy (red streak)

Halo detail of a far distant galaxy

Galaxy halo
There is a cloud of stars around the Andromeda galaxy. Scientists using the Hubble have measured the age of this "halo" and found that around one-third of the stars are young and only formed 6 to 8 billion years ago.

ANDROMEDA

SPIRAL GALAXY

Andromeda is a big bully. More than twice the size of the Milky Way—it was once thought to mark that galaxy's edge—it holds a trillion stars and is pulling in satellite galaxies as it hurtles at 75 million miles (120 million km) a second toward . . . us! It will arrive in four billion years. There is a massive black hole at its center and at least 20 more black holes dotted around within it.

Also called: M31 or NGC 224
Type: spiral
Constellation: Andromeda
Group: Local Group
Diameter: 260,000 light years
Number of stars: 1 trillion

Main sequence star

Most of their lives, stars continue to radiate energy because of the fusion of hydrogen. Ninety percent of the stars in the Universe are these main sequence stars. Our Sun (above) is classified as a G2 yellow dwarf star in the main sequence phase of its life.

After **10 billion years** as a **main sequence** star, our **Sun** will become a **red giant**.

Red giant

When a small main sequence star runs out of hydrogen to fuse, the core begins to shrink, and inward-falling material is converted to heat. Hydrogen fusion begins again as the heat flows outward, heating up layers of the star farther out and causing them to expand. As the heat spreads out, the temperature cools and the star glows redder as a massive red giant.

Star factory

New stars are born in vast clouds of gas and dust like the Carina nebula (right), 7,500 light years away from Earth. Matter and gas clump together to form a hard, dense core surrounded by a spinning disk of matter. When the center is hot enough, nuclear fusion begins and a star is born.

Planetary nebula

The core temperature of a red giant increases until it is hot enough to fuse the helium created from hydrogen fusion into carbon. When the helium runs out, the fusion process stops. The star collapses under the pull of gravity, loses mass, and sheds its outer layers into space as a planetary nebula like the Helix nebula (right).

White dwarf

This very small, hot star is the last stage in the life of a small or medium main sequence star. It is the shrunken remains and has a very high density—a teaspoon of its material would weigh around 15 tons. A white dwarf cools and fades over billions of years.

STAR LIFE

STELLAR LIFE CYCLE

Every star in the sky is on a journey. It grows when gravity pulls in enough dust, hydrogen, and helium to create the conditions for nuclear fusion, which triggers millions of explosions. The heavier the star, the faster it burns, so small stars do not shine as brightly but last much longer. All stars eventually run out of fuel and die, sometimes after swelling up and kicking off a massive explosion.

"Star" in the sky
It is thought now that the Star of Bethlehem that led the Magi to the stable for the birth of Jesus in Christian tradition may have been either a planet, a comet, or a supernova.

Massive star
Some really turbulent star nurseries create stars that are 15 to 30 times as massive as most main sequence stars. This is Wolf-Rayet 124 (left), over 20 times the mass of our Sun.

Red supergiant
When a massive star runs out of hydrogen, it becomes a red supergiant. It will only last up to a million years, during which time it continues to fuse heavier and heavier elements until iron builds up in the core.

Neutron star
A neutron star may be created when giant stars die in supernovas. These are very dense, city-sized objects. A single teaspoon of neutron star material would weigh 4 billion tons.

Black hole
In more massive star remnants, gravity wins. After a supernova, the core may collapse into a single point containing all the mass of the original star. Gravity is so strong that not even light can escape.

Supernova
In a red supergiant, the process of fusing iron needs more energy than it releases. The supergiant may explode as a supernova. This is the Crab nebula, the remnants of a supernova.

A bright, colorful death

When stars die in a supernova explosion, they do so in a spectacular fashion. This is a section of a supernova remnant, NGC 6960, known as the Veil Nebula, Cygnus Loop, or Witches' Broom. A star exploded to form a nebula thousands of years ago, but its shock wave is still traveling at 1.4 million mph (2.3 million km/h). It is around 110 light-years across, and this image shows only a small part, less than 1 percent. The intertwined and colorful wisps of cooling gas, originally heated to millions of degrees, are all that remain of a star that was 20 times larger than our Sun.

STARLIGHT

STARS

Colors indicate a star's surface temperature. Cool stars burn red (often so faintly that we cannot see them) while warmer ones burn white. The most fiery, short-lived giants burn blue and are easy to spot.

Watching the stars
The color of the stars seen from Earth can also give astronomers information, including the stars' temperature, brightness, and mass. But there are some things we cannot see, for example the colors green and purple, and neutron stars emit most of their radiation in X-rays.

Blue stars
These are the brightest, hottest young stars, such as Rigel and Deneb, with a surface temperature of 53,500–107,500°F (29,725–59,725°C). Blue-white stars have surface temperatures of 17,540-53,500°F (9,725-29,725°C).

White stars
Most white stars, such as Altair, are hotter and more massive than our Sun, with a surface temperature of 13,000–17,540°F (7,225–9,725°C). The tiny, dim, white dwarfs are also in this category.

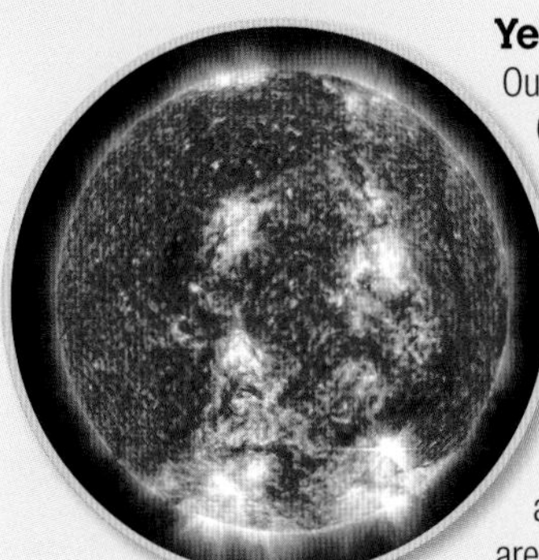

Yellow stars
Our Sun and Alpha Centauri are examples of yellow stars, with a surface temperature of 8,540–13,000°F (4,275–7,225°C). The majority of these stars are around the same size as the Sun, but there are also yellow giants.

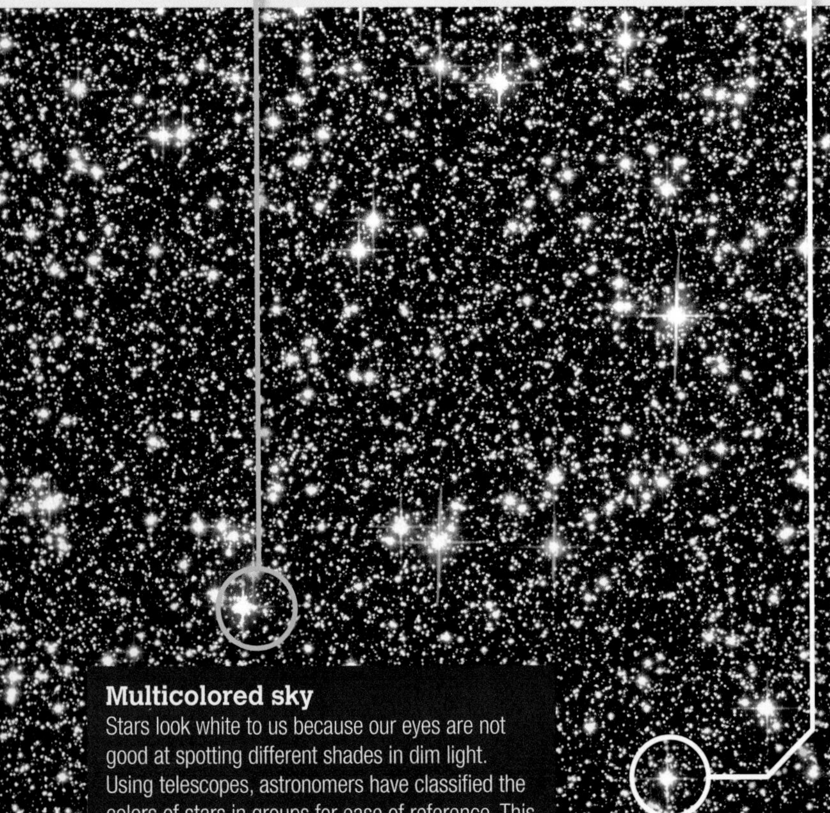

Multicolored sky
Stars look white to us because our eyes are not good at spotting different shades in dim light. Using telescopes, astronomers have classified the colors of stars in groups for ease of reference. This Hubble image of star clouds in the constellation of Sagittarius demonstrates the range of colors from the hottest (blue) to the coldest (red).

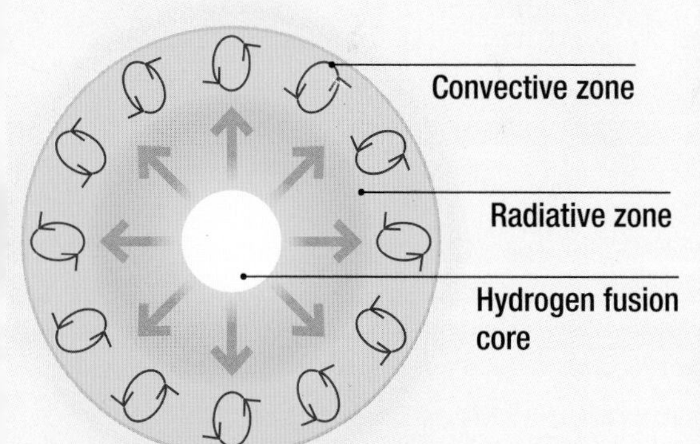

How stars work

In a small main sequence star, the extreme temperatures and pressures in the core force hydrogen nuclei (protons) together to create helium. It takes six hydrogen nuclei to fuse into each helium-4 nucleus. This process releases huge amounts of energy that take hundreds of thousands of years to reach the surface of the star and travel out into space.

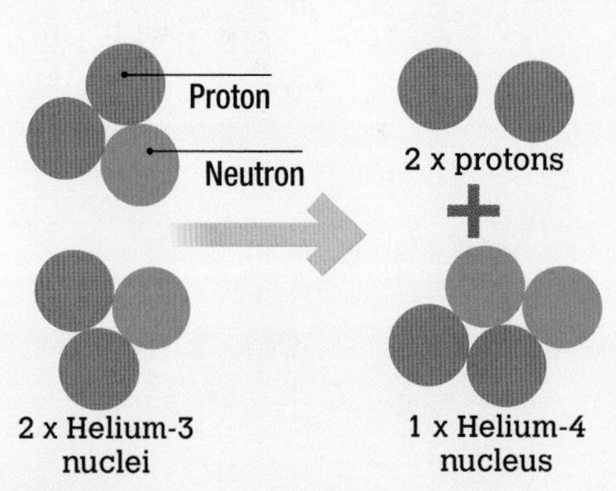

Power process

Heat and pressure inside the core of a star such as our Sun are so high that protons hit each other hard enough to bond together. If two helium-3 nuclei smash together, the result is a single helium-4 nucleus (made up of two protons and two neutrons) plus two protons. When this happens, energy is released.

Orange stars

Older, dying stars such as Cygni B and Betelgeuse are in this category, with a surface temperature of 5,480–8,540°F (3,225–4,275°C). Most are smaller than our Sun, although there are also huge orange giants that are also near the ends of their lives.

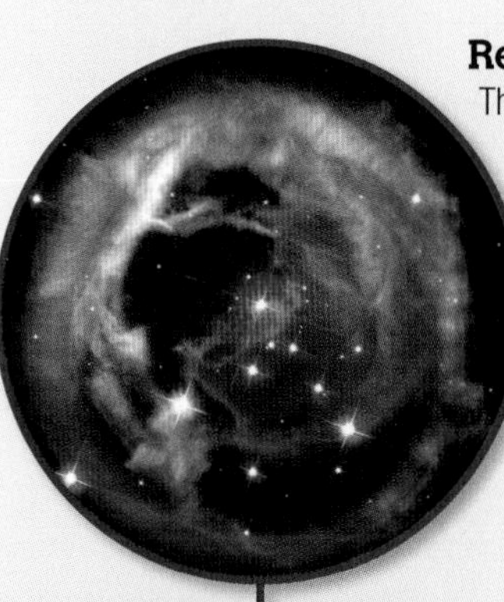

Red stars

These stars have surface temperatures of up to 5,480°F (3,225°C). The group includes enormous but very distant dying stars, the red giants, such as Mira and Antares.

We know of **220 Wolf-Rayet stars**, the **hottest** stars of all, in the **Milky Way**, but there may be many more.

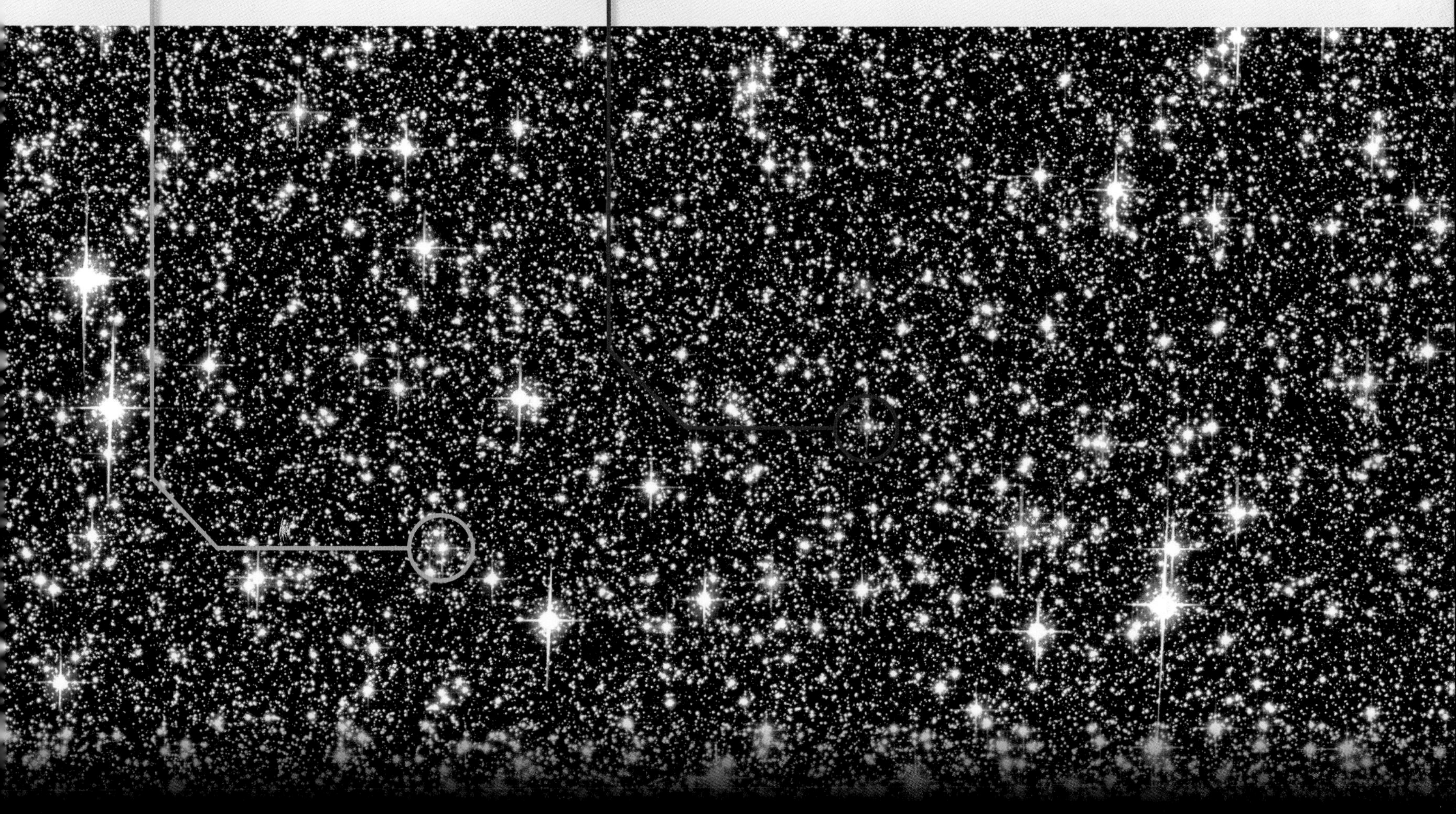

Blue supergiants

This is the Tarantula nebula in the Large Magellanic cloud, an immense star-forming region. In its center are a large number of blue supergiants that are among the most luminous stars—they can be millions of times more luminous than the Sun. Blue supergiants are short-lived and end in spectacular supernova explosions.

INCREDIBLE STARS

Everything in the night sky has a story—and sometimes a secret. Some dots of light are actually pairs of stars orbiting together. These are binary stars, and scientists believe they help create the right conditions for life. Others are all that remains of giant supernovas—these neutron stars have more mass than our Sun because they hold the collapsed matter left after a massive expansion as a star died.

Debris disks

These disks of dust and debris orbit around a star. They are the remnant of the planet-forming process, so astronomers looking for exoplanets (*see pp.190–191*) search for them. There are thousands of main sequence stars that are known to host planets or debris disks, but only a fraction of those are known to have both planets and debris disks.

Star HD 32297 with a disk of dust and debris

Binary and variable stars

Our Sun is on its own, but usually there are at least two stars orbiting each other, and these are known as binary stars. Around 80 percent of stars are binary, and the brightest of the two stars in each case is the primary. Some binary stars are variable—they change their brightness. This may happen because they are a very close pair and are exchanging mass, or because one star eclipses the other.

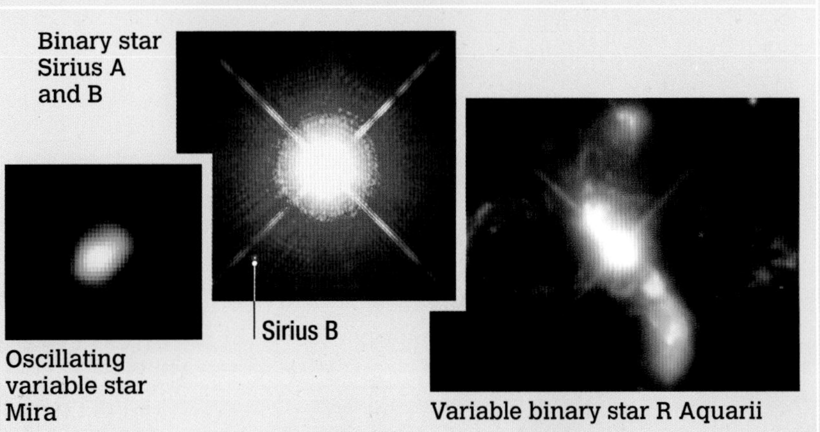

Binary star Sirius A and B

Oscillating variable star Mira

Variable binary star R Aquarii

Neutron stars

These small stars are usually up to 18 miles (30 km) across. A neutron star is formed when a massive star collapses after a supernova. Protons and electrons melt into one another to form neutrons, which are tightly packed together. Neutron stars can cause gravitational waves, ripples in spacetime (*see p.45*).

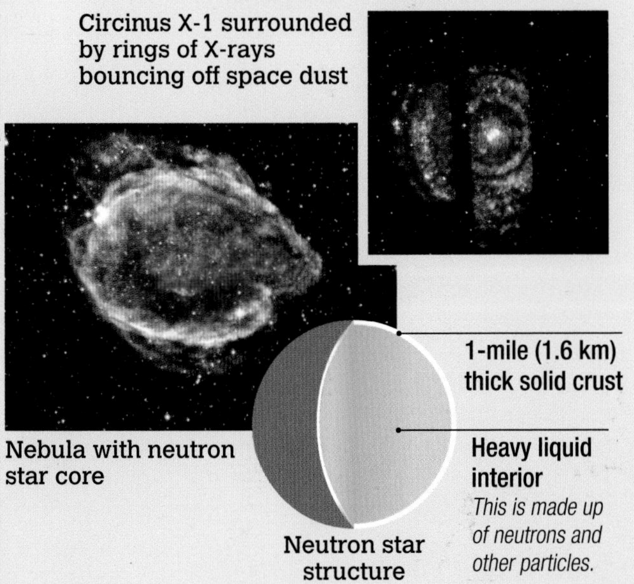

Circinus X-1 surrounded by rings of X-rays bouncing off space dust

Nebula with neutron star core

Neutron star structure

Some **neutron stars** are **magnetars**, with a very powerful **magnetic field**.

Pulsars

These neutron stars spin rapidly and produce electromagnetic radiation along a narrow beam. Their short, regular spins produce an exact interval between pulses that can only be seen when Earth is close enough to the direction of the beam.

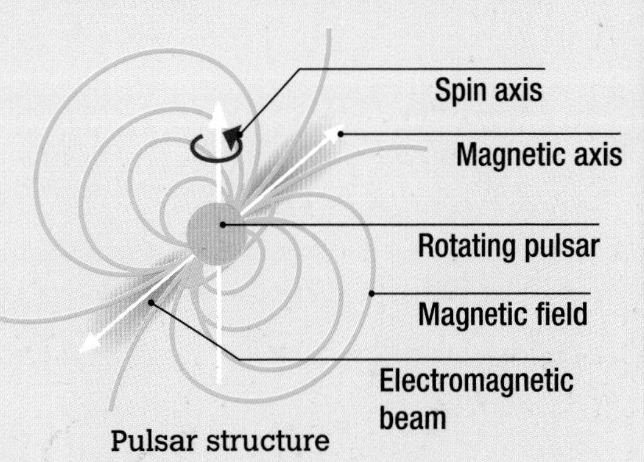

Pulsar structure

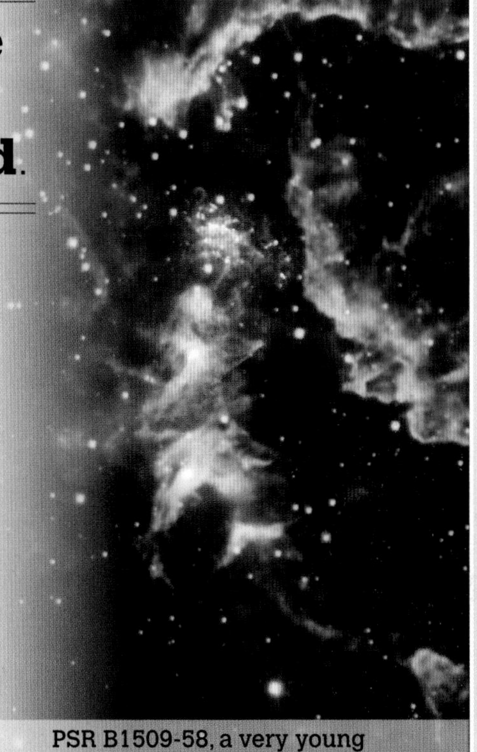

PSR B1509-58, a very young and powerful pulsar

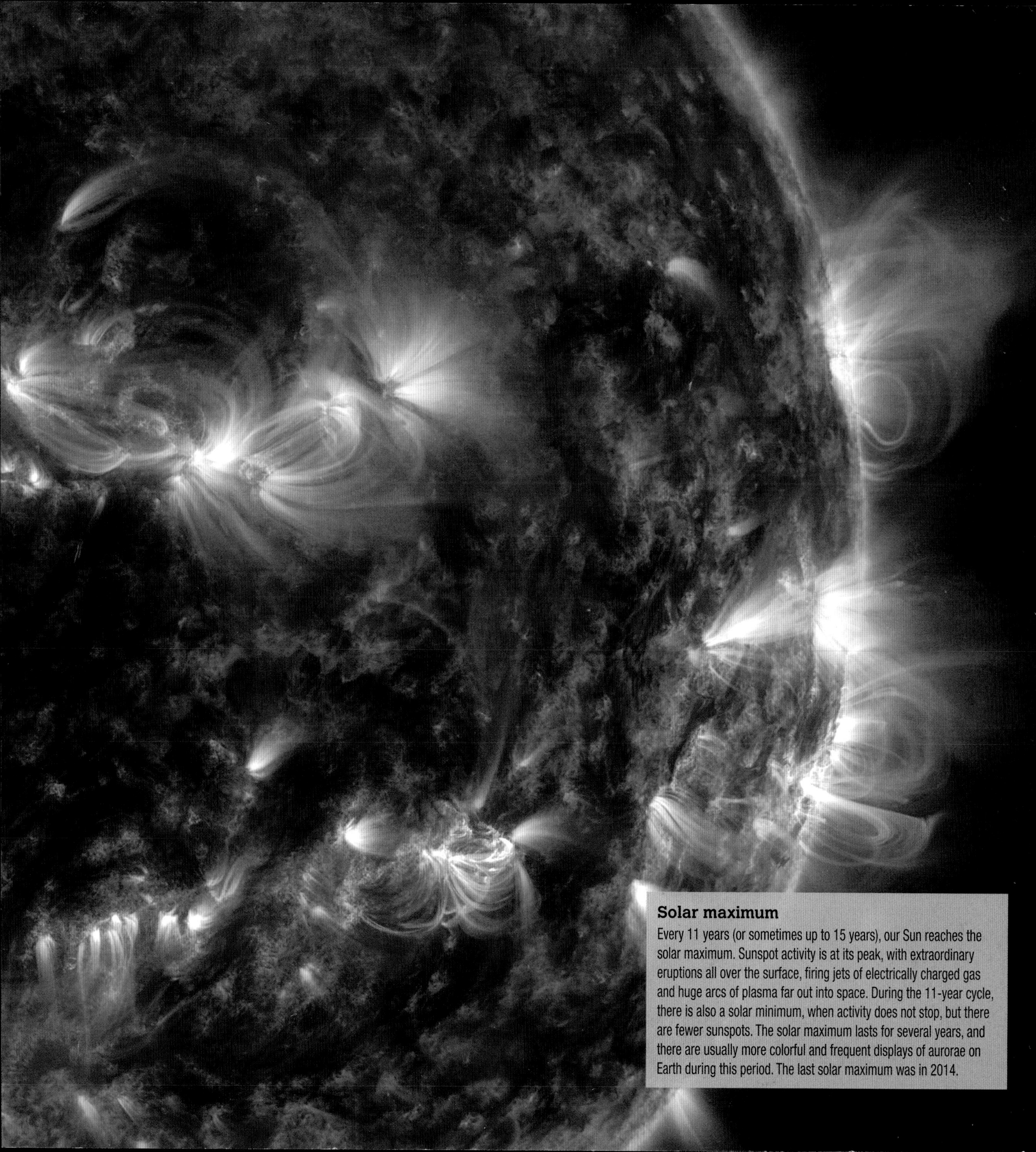

Solar maximum

Every 11 years (or sometimes up to 15 years), our Sun reaches the solar maximum. Sunspot activity is at its peak, with extraordinary eruptions all over the surface, firing jets of electrically charged gas and huge arcs of plasma far out into space. During the 11-year cycle, there is also a solar minimum, when activity does not stop, but there are fewer sunspots. The solar maximum lasts for several years, and there are usually more colorful and frequent displays of aurorae on Earth during this period. The last solar maximum was in 2014.

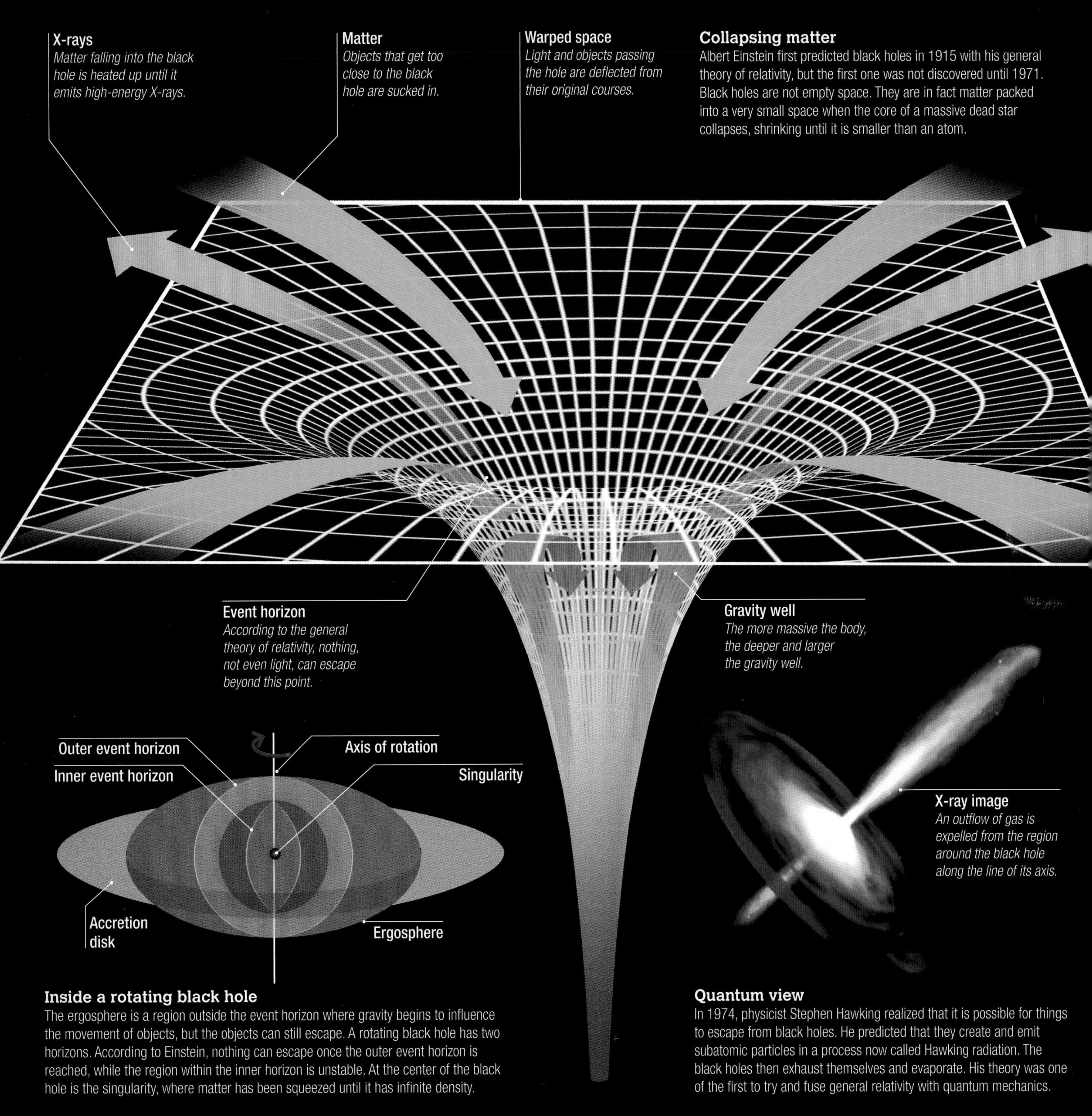

Collapsing matter

Albert Einstein first predicted black holes in 1915 with his general theory of relativity, but the first one was not discovered until 1971. Black holes are not empty space. They are in fact matter packed into a very small space when the core of a massive dead star collapses, shrinking until it is smaller than an atom.

Inside a rotating black hole

The ergosphere is a region outside the event horizon where gravity begins to influence the movement of objects, but the objects can still escape. A rotating black hole has two horizons. According to Einstein, nothing can escape once the outer event horizon is reached, while the region within the inner horizon is unstable. At the center of the black hole is the singularity, where matter has been squeezed until it has infinite density.

Quantum view

In 1974, physicist Stephen Hawking realized that it is possible for things to escape from black holes. He predicted that they create and emit subatomic particles in a process now called Hawking radiation. The black holes then exhaust themselves and evaporate. His theory was one of the first to try and fuse general relativity with quantum mechanics.

BLACK HOLES

SPACE SCIENCE

Spaghettification is a great word for a mysterious and powerful event—the moment when an object is pulled by such formidable gravity that it stretches out like a length of pasta as it is pulled down and down. That is what black holes do. A dying star collapses and its dense mass concentrates into such strong gravity that, according to Einstein's general theory of relativity (*see p.45*), nothing escapes once it reaches the event horizon, not even light.

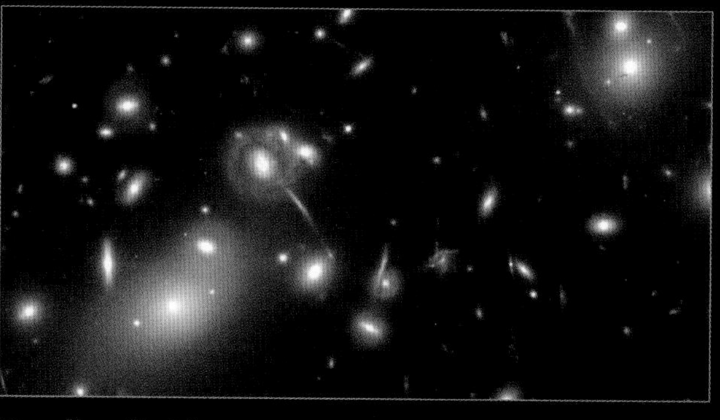

Bending light
The gravity of black holes bends light that passes close but is not pulled into them. This forms halos around the black holes. In this image of the Abell 2218 cluster, the galaxy is bending light from more distant galaxies behind it. This creates the thin arcs of light that are visible.

The black hole **Cygnus X-1** spins at **more than 800 times** a second.

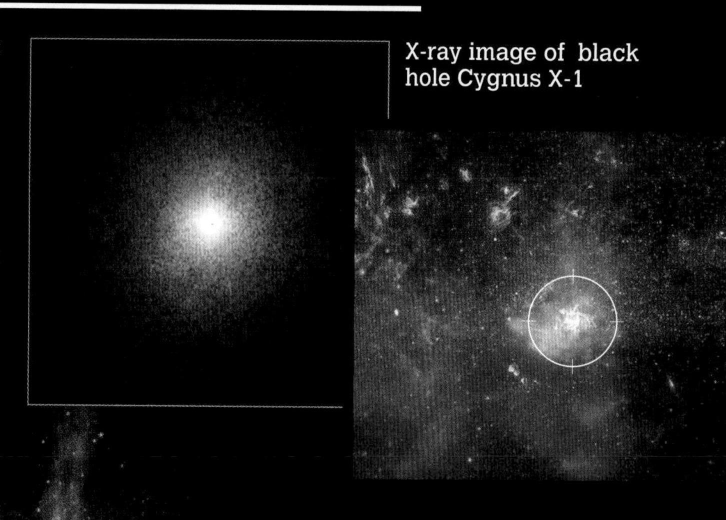

X-ray image of black hole Cygnus X-1

Black hole in the Milky Way

Contrast in mass
Cygnus X-1, the first black hole to be identified in 1971, is 6,070 light-years away from Earth near large star-forming regions in the Milky Way. Scientists have established that it is 14.8 times the mass of our Sun. In 2017, a black hole that is 100,000 times the mass of our Sun was found in a gas cloud near the center of the Milky Way.

Visible proof
The event causes X-rays to arise that are seen for years.

Star material
The cloud of debris from the star is blown away from the black hole.

Tidal disruption
If a star passes close to a supermassive black hole, tidal forces rip it apart and drag a stream of debris into the hole while blowing the rest away into space. In 2014, astronomers observed an eruption of optical, ultraviolet, and X-ray light from ASASS N-14li, in the center of a galaxy 290 million light-years away from Earth.

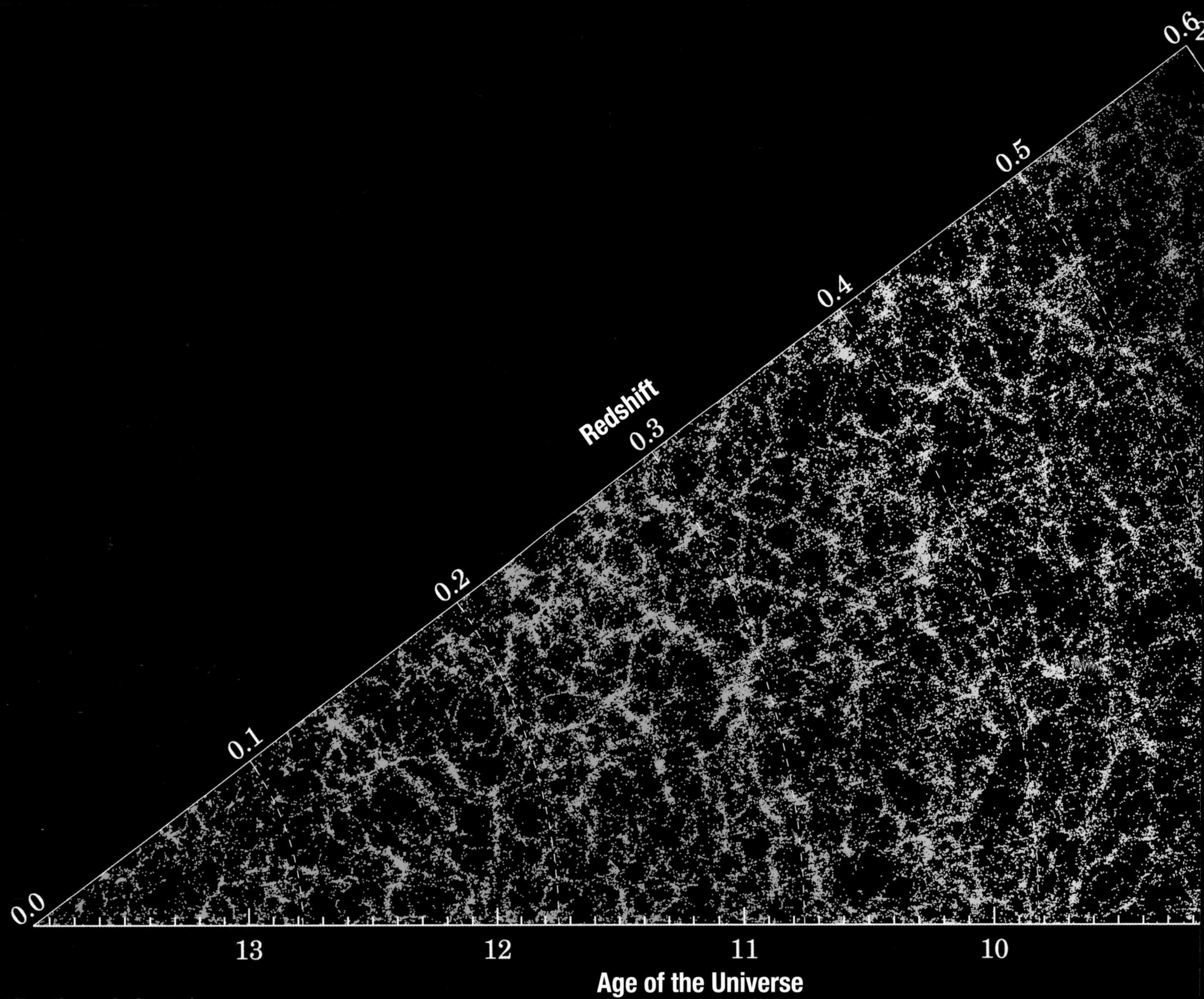

Slice through the Universe

When we look at maps, we often look at slices through the Universe like this one based on observations made with the 6.5-meter telescope at the MMT Observatory in Arizona. There are vast empty regions called voids that contain few if any galaxies. These voids are surrounded by thin walls and filaments where the galaxies cluster together. Thousands of galaxies mark the perimeter of the biggest voids. Here, galaxies trace the cosmic web from the present time (left) to the middle-aged Universe, an era six billion years ago, (right) measured by the galaxy redshift *(see p.189)*.

240
230
R.A. (deg)
220
210
200

MAPPING THE UNIVERSE

BY DR. MARGARET J. GELLER, ASTROPHYSICIST, HARVARD–SMITHSONIAN CENTER FOR ASTROPHYSICS

A map of the Universe is a journey of the imagination. It reveals the largest patterns we know in nature. It is a map in space and time because when we look out in space, we look back in time. As we look to larger and larger distances we see the Universe as it once was.

The early Universe was so hot that hydrogen was ionized (split into electrically charged particles) and the ancient photons (particles of light) could not travel freely. When the Universe became less dense and cooler at the age of 400,000 years, the photons could travel freely. We observe the lumps and bumps in the distribution of matter in the early Universe as small differences in temperature in the microwave background radiation in different regions of space.

In our puzzling Universe, 84 percent of this matter is dark. We do not know what it is, but we know a lot about where it is—we can use the galaxies we see to trace dark matter. Over 14 billion years, gravity has amplified the small lumps and bumps to make the beautiful patterns that make up the cosmic web. I played a central role in the discovery of these patterns by mapping the nearby Universe. Today, our maps reach into the middle-aged Universe and earlier.

Astrophysicist
Dr. Geller works at the Harvard-Smithsonian Center for Astrophysics. She is a pioneer in mapping the Universe. She leads large projects that map the dark matter in the Universe.

To make the maps, we begin with pictures of the sky that give us the latitudes and longitudes of the galaxies. We measure distance by spreading out the light from each galaxy into its colors. We make measurements using complex instruments on large telescopes. As a result of the expansion of the Universe, the wavelength of light traveling toward Earth through space also stretches and the light becomes redder. This redshift tells us the distance to the galaxy and the time when the light left the galaxy. We look back in time when we look out in space.

We can understand how the patterns in the Universe change and the way galaxies trace the dark matter by comparing our maps with simulated maps made on the world's largest computers—the 14-billion-year history of the Universe is simulated in only a few months. We then compare the patterns they contain with the patterns we observe.

"My **goal** is to **understand** how the **Universe** looks **today** and how it looked **billions of years ago**."

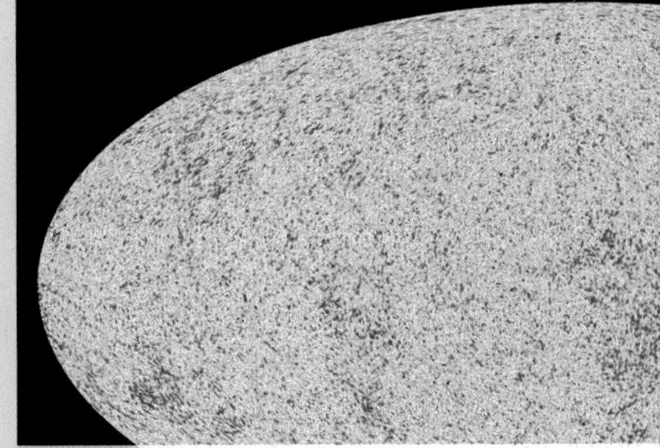

Map of the 400,000-year-old Universe showing small temperature differences in different areas of space

Dark matter showing the smooth nature of the 200-million-year-old Universe

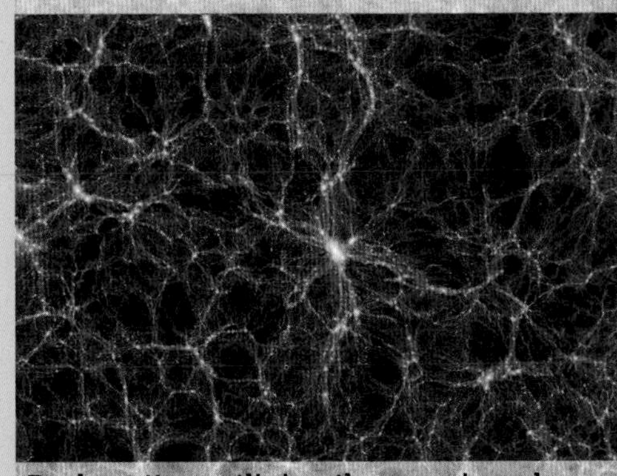

Dark matter outlining the cosmic web in today's lumpy Universe

Cluster of galaxies about 4 billion light years from Earth

COSMIC COLLISIONS

MERGING GALAXIES

Everything in space is moving, so sometimes things collide. There might be a huge smash or a more gentle merging as gravity changes the pattern to make new shapes. But the friction generated as gas and dust shift around raises temperatures, and when there is sufficient mass, huge explosions can create new stars. And if neutron stars meet, the results are sure to be spectacular . . .

MULTIPLE GALAXIES

SPIRAL DANCE

When several galaxies are attracted to one another, their movement is like a strange, slow dance as they spiral around each other. Eventually they will combine into a giant.

Merging galaxies
Stephan's Quintet is a group of five galaxies, although NGC 7320 is actually seven times closer to Earth than the rest of the group. Three of the others show elongated shapes and distortions because of close encounters with each other.

NGC 7320

NGC 7319

NGC 7318A

NGC 7318B

NGC 7317

Fairytale formation
Some multiple galaxies spiral around each other for billions of years, turning into swirling trails of stars. The Tinker Bell Triplet (left), 650 million light-years away from Earth, is the result of the coming together of three galaxies.

NEUTRON CRASH

ANCIENT COLLISION

Neutron stars hold massive amounts of energy in their dense cores. In 2017, astronomers picked up the sound of two crashing 130 million years before and saw a gigantic explosion whose heat created heavy elements.

Bright flash
The supernova created by the collision of the two neutron stars resulted in ripples in spacetime (*see p.45*), as well as light, which was released in a two-second gamma-ray burst. The equivalent of the mass of 28 Jupiters was converted into energy.

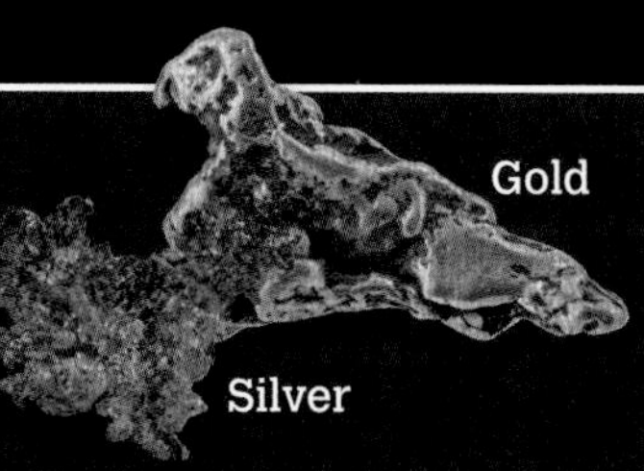

Heavy metal-maker
The collision created heavy elements including silver, iron, and lead. It created ten times the weight of Earth in gold and platinum alone.

DUAL GALAXIES

PAIRINGS IN SPACE

Most galaxies are so spread out that when they meet they tend not to collide so much as mix together, pulled about by the different gravitational forces. Some parts might spin off, but the result is a larger galaxy with a different structure that is even more powerful and able to absorb any other galaxy that comes within range.

Stolen stars

The Milky Way's 11 most distant stars are around 300,000 light-years away from Earth. Astronomers think, because of their position and velocities, that five of these stars have been pulled away from the Sagittarius dwarf galaxy (left).

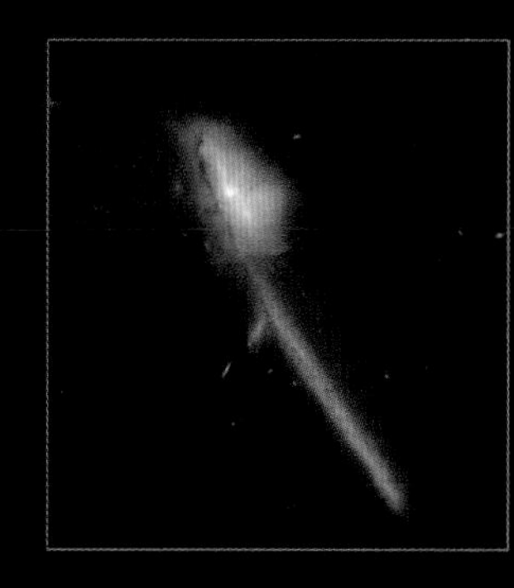

Long tail

Markarian 273 looks rather like a toothbrush. Its 130,000 light-year-long tail and nucleus with two components indicate that it is the result of a merger between two galaxies. It is a young star-forming galaxy.

Starburst

This pair of merging galaxies are known as the Antennae galaxies. When galaxies collide the event can cause huge amounts of star formation. The Antennae galaxies began their interaction several hundred million years ago, and are currently in the starburst phase.

Space punctuation

Astronomers call this collision that is 450 million light-years away from Earth the "cosmic exclamation point." Arp 302 is made up of two colliding galaxies that are spinning into one another. They will finally merge millions of years from now.

Two tails, two hearts

NGC 3256 is part of the Hydra-Centaurus supercluster around 100 million light-years away from Earth. It has two luminous tails swirling out that indicate that it is two galaxies that are slowly colliding, and it has two distinct nuclei in its center.

Cosmic rose

These two interacting galaxies are 300 million light-years away from Earth. To many, Arp 273 looks like the flower and stem of a rose. The galaxies are separated by tens of thousands of light-years from each other, but the disk of UGC 1810 (right) is distorted by the gravitational pull of its smaller companion galaxy UGC 1813 (left) which may have dived through the larger galaxy. UGC 1813 is edge-on in this image and shows signs of starburst activity—star formation—in its nucleus, which is probably caused by the close encounter.

EXPANDING UNIVERSE

SPACE SCIENCE

In 1929, Edwin Hubble proved there are galaxies other than ours, and they are all moving. Scientists assumed this movement was gradually slowing down after the birth of the Universe. However, in 1998, astronomers watching distant supernovas realized that these stars were traveling farther away from Earth and that this showed that the Universe is still expanding—and at an ever-faster rate.

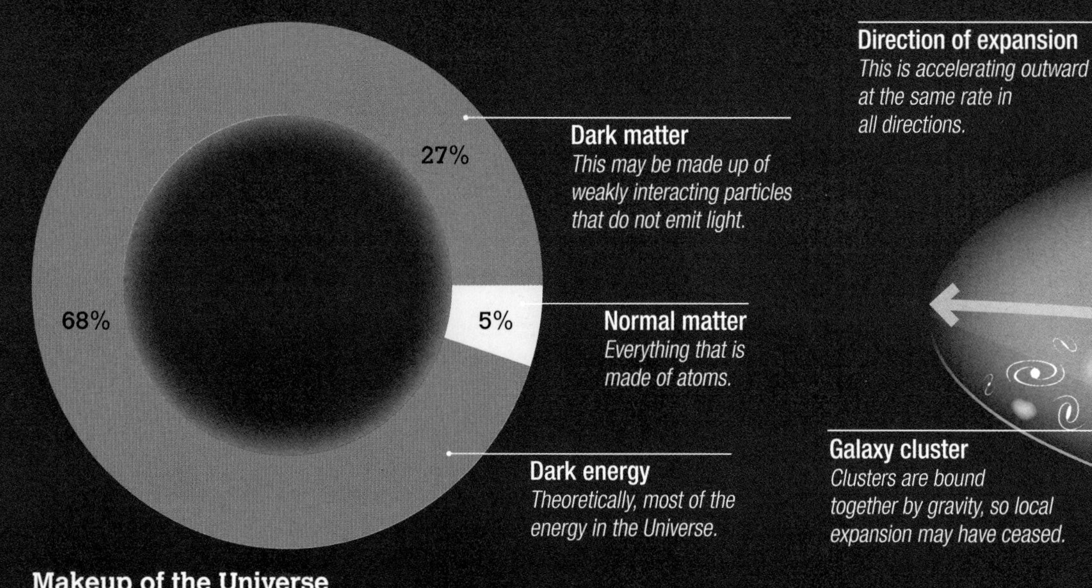

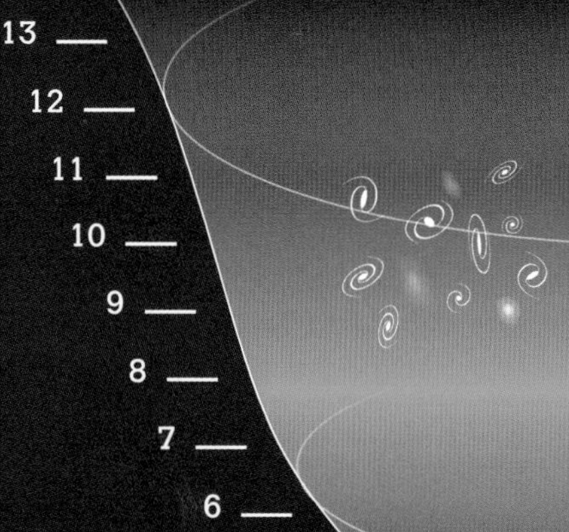

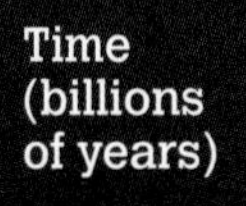

Makeup of the Universe

Scientists have found that there is not enough gravity being generated by observable matter in the Universe to keep everything in its place, so there must also be a mysterious force that we cannot see—dark matter. Dark energy is thought to act in opposition to the Universe's gravity and may be causing its expansion to accelerate.

Dark matter observed

The red-tinted clumps here are a map of the dark matter in the Abell 901/902 supercluster. To make the map, astronomers analyzed the effects of gravitational lensing, where light from more than 60,000 galaxies behind the cluster is distorted by intervening matter.

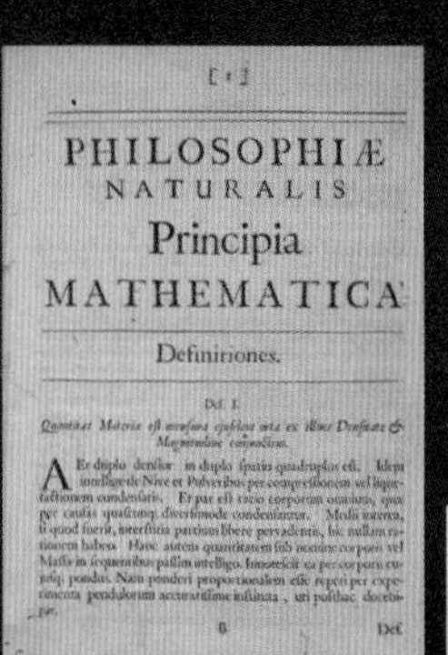

PHILOSOPHIÆ
NATURALIS
Principia
MATHEMATICA

Definitiones.

Issac Newton's *Principia*

Universal theories

Isaac Newton's model for a static stellar system offered in his *Principia*, first published in 1687, held sway for centuries. It took Edwin Hubble and improved telescopes to demonstrate there was something outside our galaxy—a whole Universe that was growing in size.

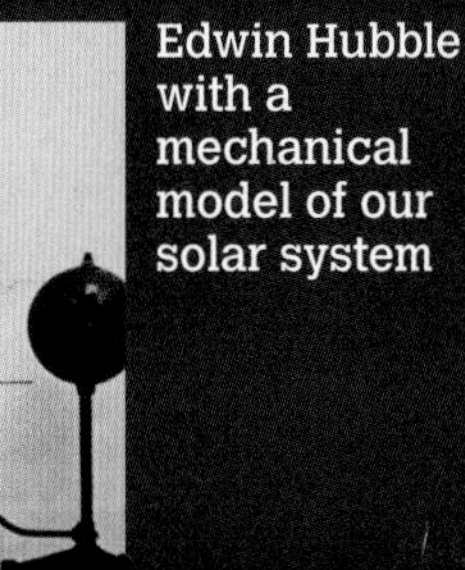

Edwin Hubble with a mechanical model of our solar system

Dark matter cannot be seen by **using** a **telescope**: it does not **emit** or **absorb** light.

Distant galaxies
Before Edwin Hubble, it had not been realized that the Universe extended far beyond the Milky Way. The Hubble telescope allows astronomers to see distant galaxies far more clearly than any telescope before and measure their movements using redshift (right). This is an image from the survey Hubble Ultra Deep Field.

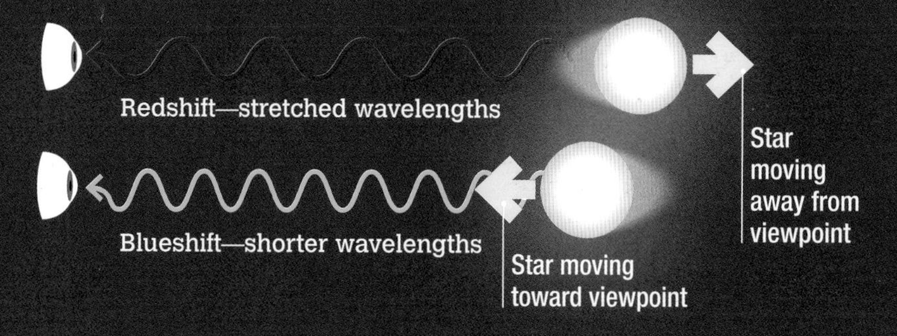

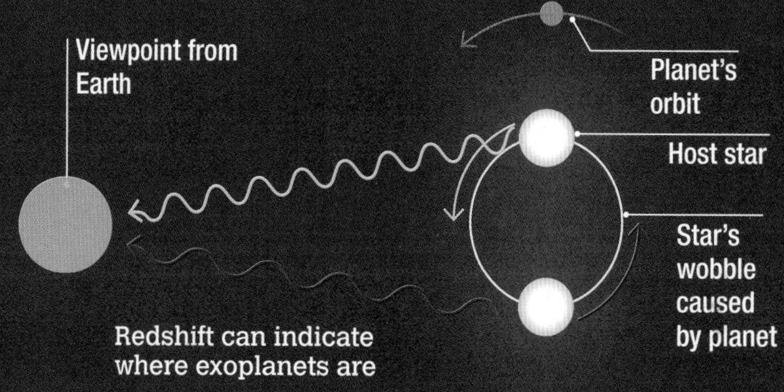

Redshifting

The movement of stars and galaxies away from Earth can be calculated. The movement affects the amount of light that we receive from an object in space. If the object is moving away, astronomers can see that the wavelength of the light is stretched. In other words the light is "shifted" to longer wavelengths, toward the red part of the spectrum—it is "redshifted." The amount it has redshifted can be measured by observing the subtle movement of tiny, dark lines in a star's spectrum.

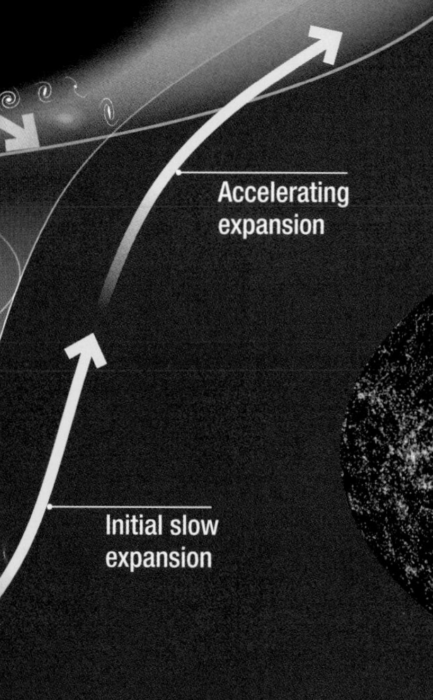

Accelerating rate of expansion

Edwin Hubble suggested that, since the Universe's possible formation 13.8 billion years ago, it has been getting ever larger. He determined that galaxies are not moving through space, rather they are moving in space because the space itself is moving, and the Universe has no center because everything is moving away from everything else. The expansion rate of the Universe was dubbed the Hubble Constant.

Redshifting the Universe

Hubble matched up the redshifts of certain galaxies to estimated distances and discovered that the farther away a galaxy is, the faster it is moving away. Even if both objects were stationary in space, there would still be a redshift if the space between them were expanding. This led him to the understanding that the Universe is expanding uniformly. Redshifts are now part of the basic toolkit used by astronomers in their observations.

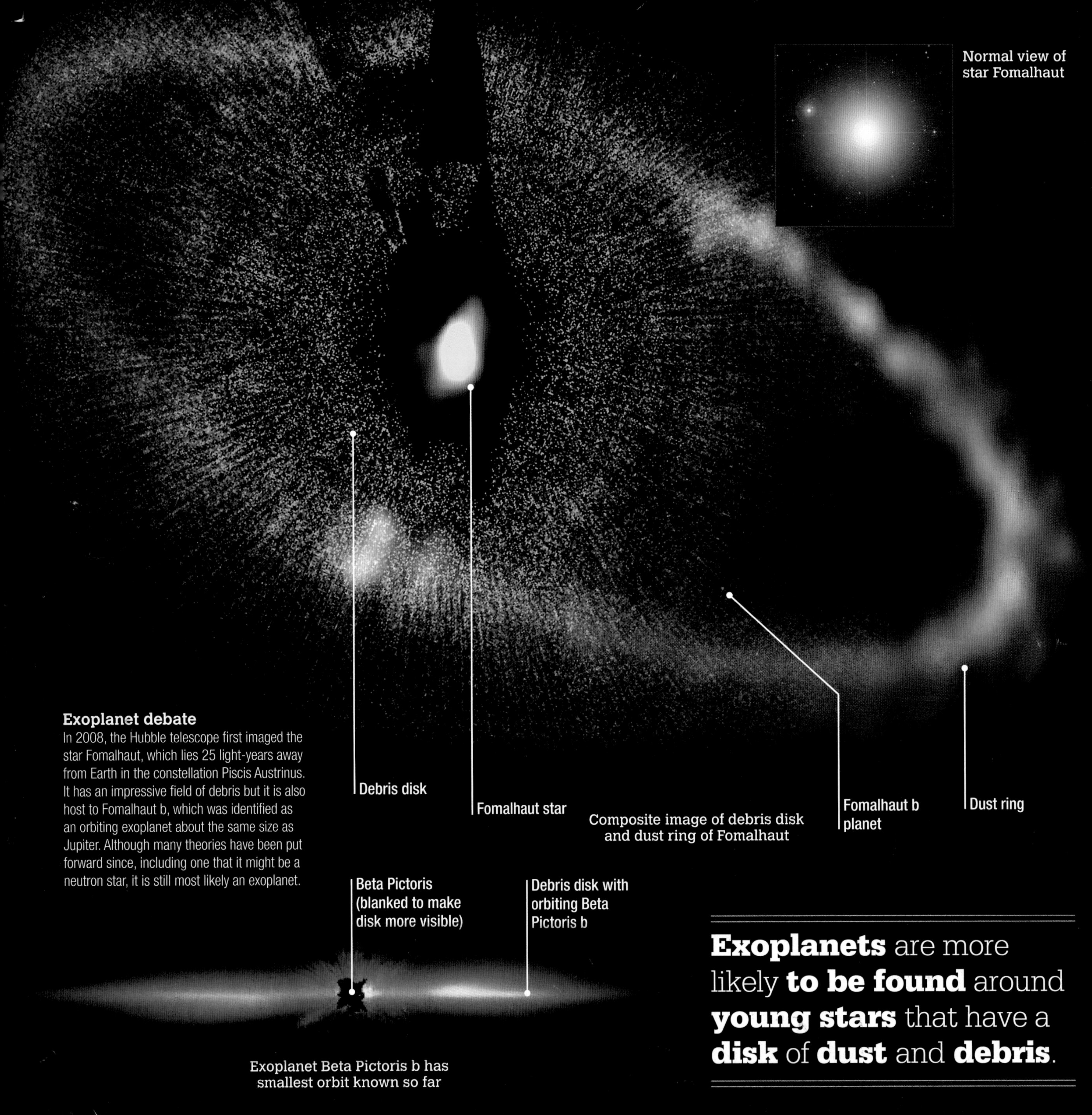

Composite image of debris disk and dust ring of Fomalhaut

Exoplanet Beta Pictoris b has smallest orbit known so far

Exoplanet debate

In 2008, the Hubble telescope first imaged the star Fomalhaut, which lies 25 light-years away from Earth in the constellation Piscis Austrinus. It has an impressive field of debris but it is also host to Fomalhaut b, which was identified as an orbiting exoplanet about the same size as Jupiter. Although many theories have been put forward since, including one that it might be a neutron star, it is still most likely an exoplanet.

Exoplanets are more likely **to be found** around **young stars** that have a **disk** of **dust** and **debris**.

EXOPLANETS

LANETS

: would take many lifetimes to even reach the parts of space that cientists are now studying in search of exoplanets—planets orbiting star that is not our Sun. Most of these are too far away to see, but stronomers find evidence of them in changes of light or the path of tars. Thousands of exoplanets have been identified, and there are robably billions out there. Some are gas giants, others appear to ave rocks and water. Are any of them able to support life like Earth?

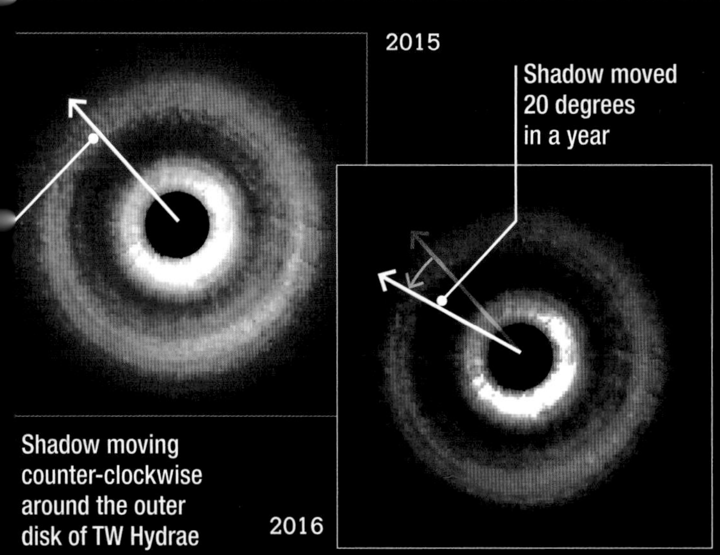

Shadow moving counter-clockwise around the outer disk of TW Hydrae

Shadow observation

These two pictures of the young star TW Hydrae, 176 light-years away from Earth in the constellation Hydra, were taken a year apart by the Hubble space telescope. Astronomers have concluded that there is a planet pulling on material near the star and warping the inner part of the disk. The shadow that is moving counter-clockwise around the outer disk is not caused by the planet itself, but rather by the warped inner disk.

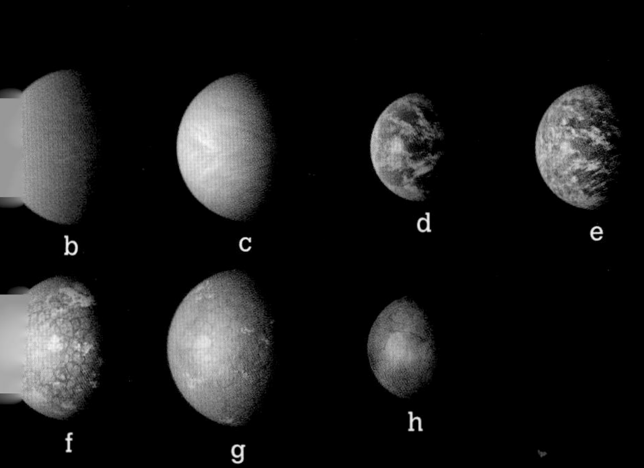

rappist 1

is planetary system is only 39 light years away from Earth in the nstellation of Aquarius. Its name comes from the telescope that ade the initial discovery in 2016. Seven planets have been identified biting around a star that is slightly larger than Jupiter. The planets e around the sizes of Venus and Earth and are all rocky

Survey first

The Transiting Exoplanet Survey Satellite (TESS) will spend two years hunting for exoplanets orbiting the brightest stars just outside our solar system. It will be surveying more than 200,000 stars for drops in brightness caused by transits

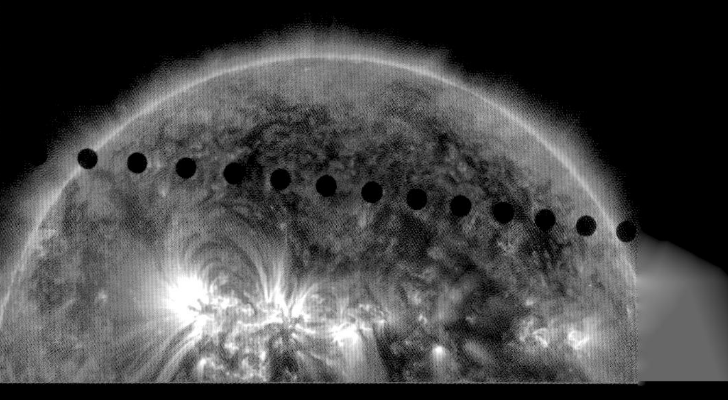

Searching for shadows

When something transits in front of a star (above), the light output is dimmed. If this happens regularly, this indicates that there is a planet orbiting that star. The change in light output may only be tiny, but it is enough to show astronomers that there is an exoplanet.

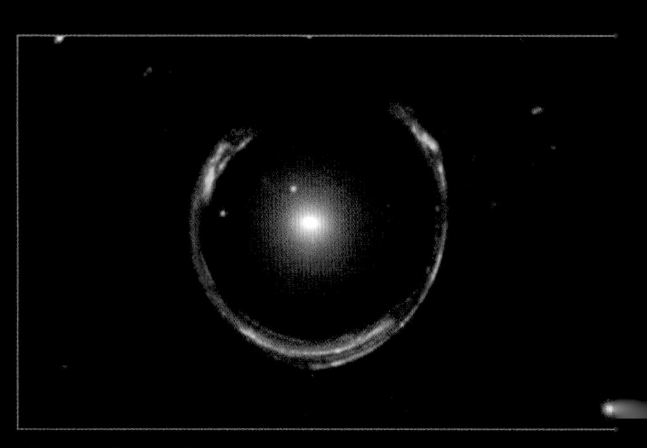

Gravitational lensing

This happens when light rays coming to the viewer from a distant bright object bend round a massive body such as a planet, massive galaxy, or black hole in their path. The presence of matter can curve spacetime, so the path of a light ray is deflected as a result—one of the predictions made by Einstein in his General Theory of Relativity (*see p.45*). In this image, light from a more distant blue galaxy is distorted and magnified by the red galaxy between it and the viewer.

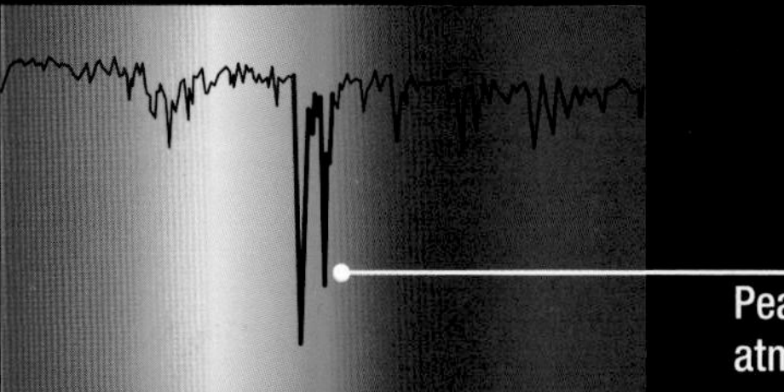

Spectroscopy

Astronomers use spectroscopy to measure the intensity of light in different wavelengths in order to find out the composition of exoplanet atmospheres. For example, if a planet has hydrogen in its atmosphere, there would an "absorption line" in a particular place on the spectra

CONTACT?

LIFE ON OTHER PLANETS

When we look at the night sky, it is natural to wonder if there is life out there in space, and there have been many so-called "sightings" of UFOs and tales of alien abduction! Today, we are developing increasingly technological ways to search for extraterrestrial life. We can look for alien signals with optical, infrared, or radio waves. We can also send probes and ultimately crewed spaceships to search for life on other planets.

Alien invasion

In the 17th century, the German astronomer Johannes Kepler wrote *Somnium* (*Dream*), a story about a journey to the Moon to meet reptile-like creatures. Since then, there has been a rich outpouring of stories, television series, and movies with imagined aliens of all shapes and sizes, many humanoid or animal-like, but also robots made of metal.

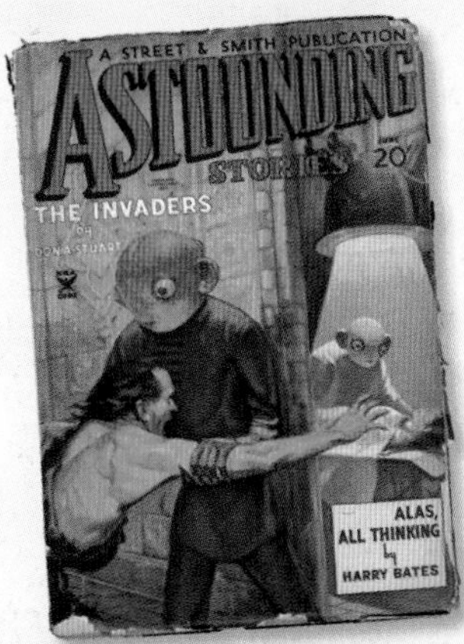

Aliens in 1935 comic book

Goldilocks zones

Earth's distance from the Sun means that it is the right temperature to support liquid water, the key ingredient for life on planet Earth. We are looking for similar conditions in other solar systems to see if life is there. Scientists call such regions "Goldilocks zones" or "habitable zones." In 2017, astronomers announced the discovery of Kepler-186f, a rocky world just 10 percent bigger than Earth, in the habitable zone of Kepler-186, a star smaller and dimmer than our Sun. Could there be life there?

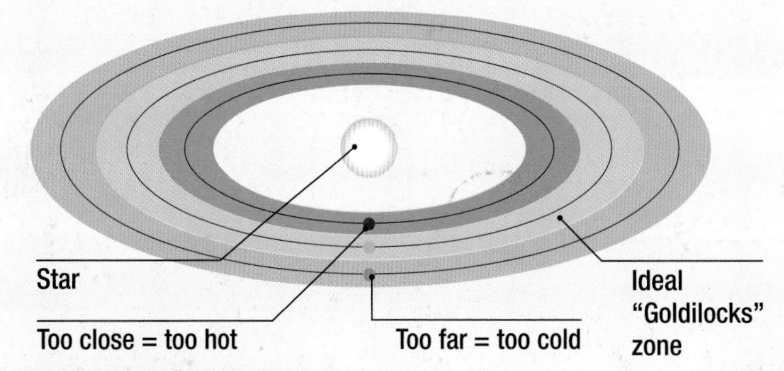

The Voyager Golden Record

The farthest of all space travelers, the Voyager probes, are carrying Golden Records of sounds and images of life on Earth. Their covers (right) have diagrams on them to help other lifeforms access the information.

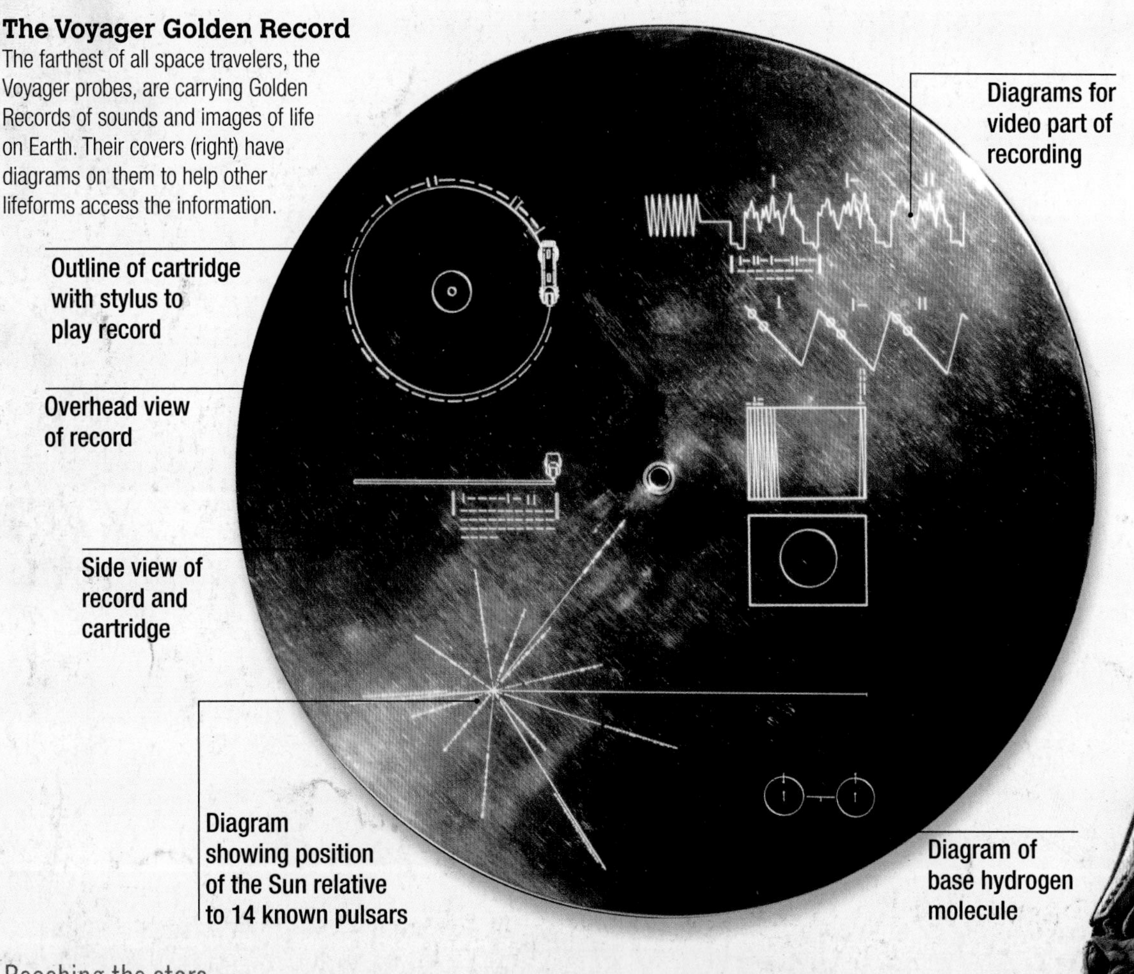

Alien from the film *Alien 3*, 1992

Alternative lifeforms

It could well be that we have not yet imagined what form life might take on other planets. In our deepest oceans, strange creatures thrive in high temperatures near hydrothermal vents. It is thought that the oceans beneath the surfaces of Saturn's moon Enceladus and Jupiter's moon Europa are the most likely places to find life in our solar system.

Hydrothermal vents have non-Sun-based life cycles

SETI projects around the world are **looking for signals** from **intelligent life** on other **planets**.

Numbers 10–1

Atomic numbers of 5 elements

Formulae of chemical building blocks of DNA

Number of nucleotides in DNA

Structure of the DNA double helix

Population, shape, and height of humans

Structure of solar system

Shape and size of Arecibo telescope

The Arecibo message

In 1974, US astronomer Frank Drake sent a coded message from the Arecibo Observatory radio telescope in Puerto Rico. He sent it toward the globular cluster M13 in the constellation of Hercules in the hope that an alien civilization would pick it up.

Searching for life

The Allen Telescope Array (ATA), at the Hat Creek Radio Observatory in north California, is a "Large Number of Small Dishes" (LNSD) designed to make space observations and search for signals for the Search for Extraterrestrial Intelligence (SETI), an organization founded in 1984.

A MOVE TO MARS

SURVIVING ON ANOTHER WORLD

How do you fancy living on another planet? It will happen one day. But there are many things to sort out first—how to dodge deadly radiation, prevent weightlessness wrecking the human body, and establish airtight shelters and a supply of oxygen, water, and food. But surely robots should take care of most of that, so we can just enjoy the amazing views—and look back at Earth?

A third less **gravity** means that **humans will feel lighter** on **Mars.**

EARTH

Distance from the Sun
93 million miles (150 million km)

Diameter
7,926 miles (12,756 km)

Length of year
365.25 days

Length of day
23 hours 56 minutes

Gravity
2.66 ttimes that of Mars

Temperature
Average 57° F (13.9°C)

Atmosphere
Nitrogen, oxygen, argon, others

MARS

Distance from the Sun
142 million miles (228.5 million km)

Diameter
4,220 miles (6,791 km)

Length of year
687 Earth days

Length of day
24 hours 37 minutes

Gravity
0.38 times that of Earth

Temperature
Average –62.8°C (–81°F)

Atmosphere
Mostly carbon dioxide, some water vapor

Why Mars?
Mars tops the list of many scientists and entrepreneurs for Earth's first space colony. Its day length is similar to Earth's, it has lots of dry land, and it has seasons. There is also water ice. On the other hand, it is much colder than Earth, with less gravity, and an atmosphere which is mostly carbon dioxide.

SpaceX's Falcon Heavy rocket

Falcon Heavy
140,660 lbs
(63,800 kg)

Space shuttle
53,790 lbs (24,000 kg)

Proton-M
50,710 lbs (23,000 kg)

Delta IV Heavy
49,740 lbs (22,560kg)

Titan IV-B
47,800 lbs (21,680kg)

Ariane 5 ES
44,090 lbs (20,000kg)

Atlas V 551
40,810 lbs (18,510kg)

Japan H2B
36,380 lbs (16,500kg)

China LM-3B
24,690 lbs (11,200kg)

Heavy payload

On February 6, 2018, the first reusable heavy-lift spacecraft was launched. The SpaceX vehicle, Falcon Heavy, is the most powerful operational rocket by a factor of two. Its two reusable side boosters landed perfectly in sync back on Earth, while the main rocket is headed in the direction of Mars.

Short first step

In 2017, ESA announced that they are working with Russia on a project to land a robot on the Moon maybe as early as 2022. The idea is that this will give the information that will make it possible to establish a colony on the Moon in the future (left). It will be a research base that could set up a radio telescope on the far side of the Moon to provide uninterrupted communication with spaceships traveling through the solar system or future settlers on other planets.

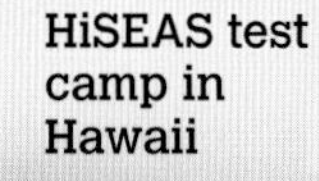

Proposed Mars One camp

HiSEAS test camp in Hawaii

Setting up camp

In the HiSEAS test camp on the volcanoes of Hawaii—the closest terrain to Mars that we have on Earth—teams spend 4–12 months at a time cut off from the world. They are learning to be self-sufficient as well as how to cope with wearing spacesuits every time they venture out of the climate-controlled pods in which they live and work. A Dutch organization, Mars One, is planning similar test sites.

Orion mission patch

Distance travel

NASA's new Orion spacecraft is being prepared to first go uncrewed beyond the Moon and return safely in the early 2020s. Then it will carry astronauts on a year-long mission to deep space as a tester for the journey to Mars.

Mars movie

The movie *The Martian*, based on a sci-fi bestseller by Andy Weir, was a huge hit because it realistically portrayed the Red Planet. NASA worked with the director on the film because this is the type of mission that they are planning.

Surviving in a barren world

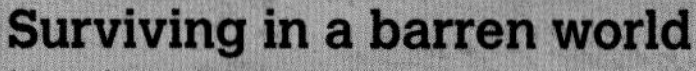

Mars is an inhospitable world, with barren red plains stretching over huge distances, and strange rock formations inside large, deep canyons. Settlers will need to set up shelters, protect themselves from radiation levels that are 250 times those on Earth, and cultivate crops—all while wearing spacesuits.

The beauty of space

The light from massive stars backdrops some of the most fantastical shapes with tall dust pillars, splashes, and blobs—as if an artist has gone mad with paint and brush. This is the Carina Nebula, 7,500 light-years away from Earth, an amazing feature in the southern hemisphere part of the Milky Way, and a tempestuous stellar nursery. Streams of hydrogen gas and dust flow out from the nebula, where fledgling stars are emerging. It is home to the explosive variable star Eta Carinae, one of the most massive stars in the Universe.

A rod or dishlike structure on spacecraft and telescopes used to transmit and receive radio signals.

aphelion
The point in the orbit of a planet, comet, or asteroid at which it is farthest from a star.

asterism
A group of stars that form a recognizable pattern but that do not form a constellation

asteroid
A chunk of rock or metal, left over from the birth of our solar system, that orbits the Sun.

astronomer
A scientist who studies the stars and other objects in space.

Astrolabe

A specialist trained to travel, live, and work in space.

atmosphere
The layer of gases that surrounds a planet or star.

aurora
A display of multicolored lights caused by particles from Earth's magnetic field dislodged by the arrival of particles from the Sun.

axis
An imaginary straight line around which an object rotates.

binary star
A system of two stars that revolve around each other, affected by each other's gravity.

black hole
The superdense collapsed core of a burned-out star that sucks in objects around it in space. Not even light can escape, so a black hole is not visible to the naked eye.

cluster
A group of stars or galaxies.

comet
A chunk of frozen gas and dust that travels in an elongated orbit around the Sun. When a comet warms up near the Sun, dust and vapor may produce spectacular "tails."

One of the many divisions of Earth's night sky described by astronomers. There are 88 constellations.

corona
The outermost part of the Sun or another star's atmosphere.

cosmic rays
Streams of high-energy, fast-moving particles that come mainly from outside our solar system.

cosmic web
A web that scientists think is composed mainly of dark matter and stretches between galaxies.

crater
A bowl-shaped depression in the surface of a planet or moon, often caused by a meteorite impact.

cryovolcano
An ice volcano, such as those found on Saturn's moon Enceladus.

dark energy
The energy that scientists believe is responsible for the accelerating expansion of the Universe.

dark matter
The invisible matter that can only be detected by observing the effect of its gravity on other objects in space.

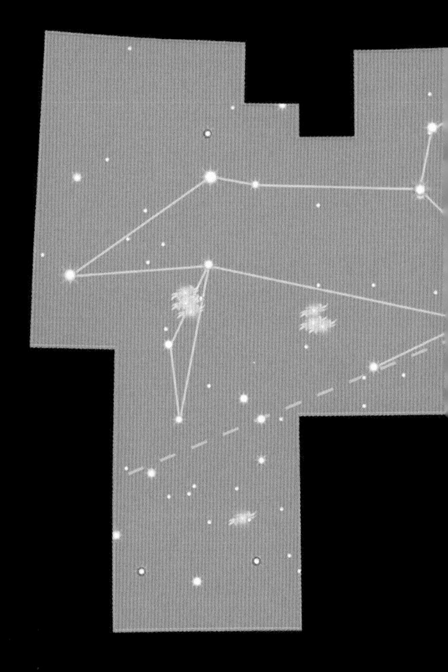
Constellation (Leo)

dwarf planet
A planet that is big enough to have become spherical but not large enough to clear other objects from its orbital path.

eclipse
An event when three objects in space line up with one another—for instance, when the Moon passes in front of the Sun, as seen from Earth.

element
A substance which is made of only one type of atom.

ESA
The European Space Agency.

EVA
An ExtraVehicularActivity, or spacewalk.

event horizon
A barrier around a black hole, this marks the point where nothing can escape the gravity of the black hole.

exoplanet
A planet that orbits a star that is not our Sun.

galaxy
A huge structure of stars, gas, and dust, usually in the shape of a spiral, an elliptical ball, or an irregular cloud.

gamma rays
An electromagnetic energy wave that has a very short wavelength.

geyser
A spring that throws out liquid water. Saturn's icy moon Enceladus has geysers.

gravitational field
The region of space around an object that is affected by the object's gravity.

gravity
The force that pulls everything toward massive objects such as planets or stars. It keeps moons in orbit around planets, and planets in orbit around the Sun.

heliosheath
The outermost layer of the heliosphere at the edge of interstellar space.

heliosphere
The place where the solar wind is slowed by the pressure of interstellar gas.

hexagonal
Having six angles and six sides.

hydrothermal
Relating to the action of water under conditions of high temperature, especially in the formation of rocks and minerals.

interplanetary
Existing between the planets of our solar system.

interstellar
Existing or occuring between stars.

infrared
A type of invisible light, produced by objects that are too cool to glow in visible light.

ionized
Describes having converted something into ions, typically by removing one or more electrons.

Comet (McNaught)

GLOSSARY

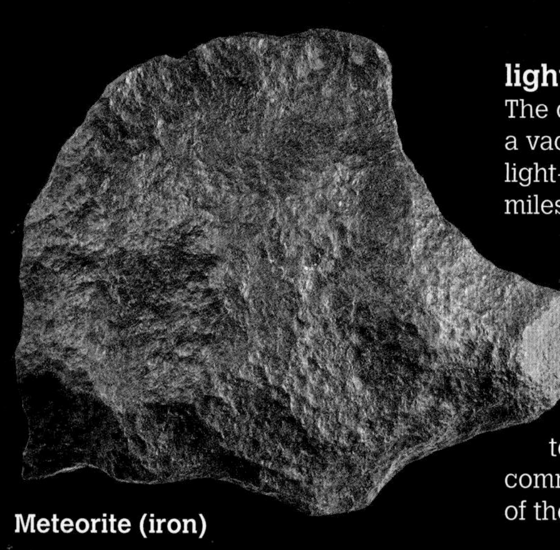
Meteorite (iron)

ISS
International Space Station.

jet stream
A current of high-speed winds.

jettison
To eject something, such as a booster rocket, from a spacecraft in flight.

lander
A space vehicle designed to land on a body in space, such as a moon or planet.

laser
An intense, focused beam of light. Laser stands for *Light Amplification by Stimulated Emission of Radiation.*

launch
To send a rocket into space.

light-year
The distance traveled by light in a vacuum in one Earth year. One light-year is equivalent to 6 trillion miles (10 trillion km).

lunar
Concerning Earth's Moon.

lunar module
A space vehicle designed to carry astronauts from the command module to the surface of the Moon and back.

magnetosphere
An area of space surrounding a planet or star that is dominated by the planet or star's magnetic field.

mare
Dark basaltic plains on the Moon. *Plural: maria*

meteor
A rock from outer space that passes through and burns up in Earth's atmosphere. Also called a shooting star.

meteorite
A fragment of rock from outer space that reaches Earth's surface instead of completely burning up in the atmosphere.

meteoroid
A rocky, metallic, or icy body traveling through space.

microgravity
Very weak gravity, as in an orbiting spacecraft.

microwave
Electromagnetic radiation with wavelengths longer than infrared and visible light, but shorter than radio waves.

MMU
Manned Maneuvering Unit, a propulsion unit used on space shuttle missions.

module
An independently operated unit that is part of the total structure of a space station such as the ISS.

molten
Melted, especially by very great heat.

moon
A solid body in orbit around a planet, a type of natural satellite.

NASA
The National Aeronautics and Space Administration.

nebula
A cloud of gas and dust in interstellar space.

neutron star
A tiny, superdense space object formed from the collapsed core of a burned-out star. It is composed of neutrons and spins very fast.

nuclear fusion
The combining of very light atomic nuclei to form a heavier nucleus. Energy is released in the process.

observatory
A building built and equipped for scientists to study objects in space.

optical telescope
A telescope that gathers and focuses light to allow us to see distant objects that we cannot see with our eyes alone.

orbit
The path taken by one object around another, larger object under the influence of its gravity.

orbiter
A spacecraft designed to orbit a planet or moon without landing.

partial eclipse
An eclipse in which one object in space is not completely obscured by the shadow or body of another.

Cargo or equipment carried into space by a rocket or spacecraft.

perihelion
The point in the orbit of a planet, comet, or asteroid at which it is closest to a star.

photon
A tiny particle of light or electromagnetic radiation.

photosphere
The luminous surface layer of the Sun or another star.

planet
A body in space that moves in orbit around a star such as the Sun. Planets reflect the light of the star that they are orbiting.

plasma
A highly energized form of gas that can carry electricity.

probe
A robotic spacecraft sent to investigate space objects up close.

A large plume of plasma, that emerges like a flame from the Sun's photosphere.

propellant
The fuel plus oxidizer that is used to propel a rocket engine.

pulsar
A neutron star with a powerful magnetic field that emits spinning beams of light.

quasar
The bright center of a galaxy, believed to be powered by an enormous black hole.

radar
A device that sends out radio waves to find out the position and speed of a moving object.

radiation
A moving electrical and magnetic disturbance that is experienced as light and heat. It is also called electromagnetic radiation.

radio wave
An electromagnetic wave that is within the range of radio frequencies.

red giant
A large, luminous star that burns helium in its core rather than hydrogen and is nearing the end of its life.

Nebula (Crab nebula)

Planet (Mars)

GLOSSARY

Spacecraft (Soyuz)

redshift
A measurement of the movement of an object in space toward longer wavelengths reflected in its shift to the red end of the spectrum.

reflecting telescope
An instrument that produces a bright, magnified image by collecting light with mirrors.

refracting telescope
An instrument that produces a bright, magnified image by collecting light with lenses.

ring system
A disk or ring orbiting an object that is composed of solid material, such as rocks, dust, or "moonlets."

rover
A vehicle for exploring the surface of an extraterrestrial body such as the Moon or Mars.

satellite
A small body that orbits a much larger one in space under the influence of the larger body's gravity. A satellite can be natural, such as a moon, or artificial.

SETI
Search for Extraterrestrial Intelligence.

solar flare
A sudden brightening in the Sun's atmosphere, caused by changes in the Sun's magnetic field.

solar system
Everything that is held in orbit around a star, such as the Sun, by its gravity.

solar wind
A stream of particles that blows across our solar system from the surface of the Sun.

spacewalk
A period of activity spent outside a spacecraft by an astronaut, usually to conduct repairs or install equipment.

star
A huge, glowing ball of gas that gives off its own heat and light. The Sun is a star.

sunspot
A region of intense magnetic activity in the Sun's photosphere that appears darker than anything around it.

supernova
A huge explosion that takes place when a star has used up all of its fuel and collapses.

termination shock
The boundary marking one of the outer limits of the Sun's influence in interstellar space.

terrae
Highlands on the Moon.

thrust
The force of an engine that propels a rocket or spacecraft forward.

transit
The passage of a planet or star in front of another, larger object.

UFO
Unidentified Flying Object.

ultraviolet
A type of invisible light, produced by objects that are too hot to glow in visible light.

vacuum
An empty space in which there is no air or other gas.

white dwarf
A hot, dense space object, formed by the core of a burned-out star.

X-ray
A form of radiation produced by stars and hot gas clouds. X-rays are high-frequency waves emitted by some of the most violent processes in the Universe.

Star field

GLOSSARY

Aristotle

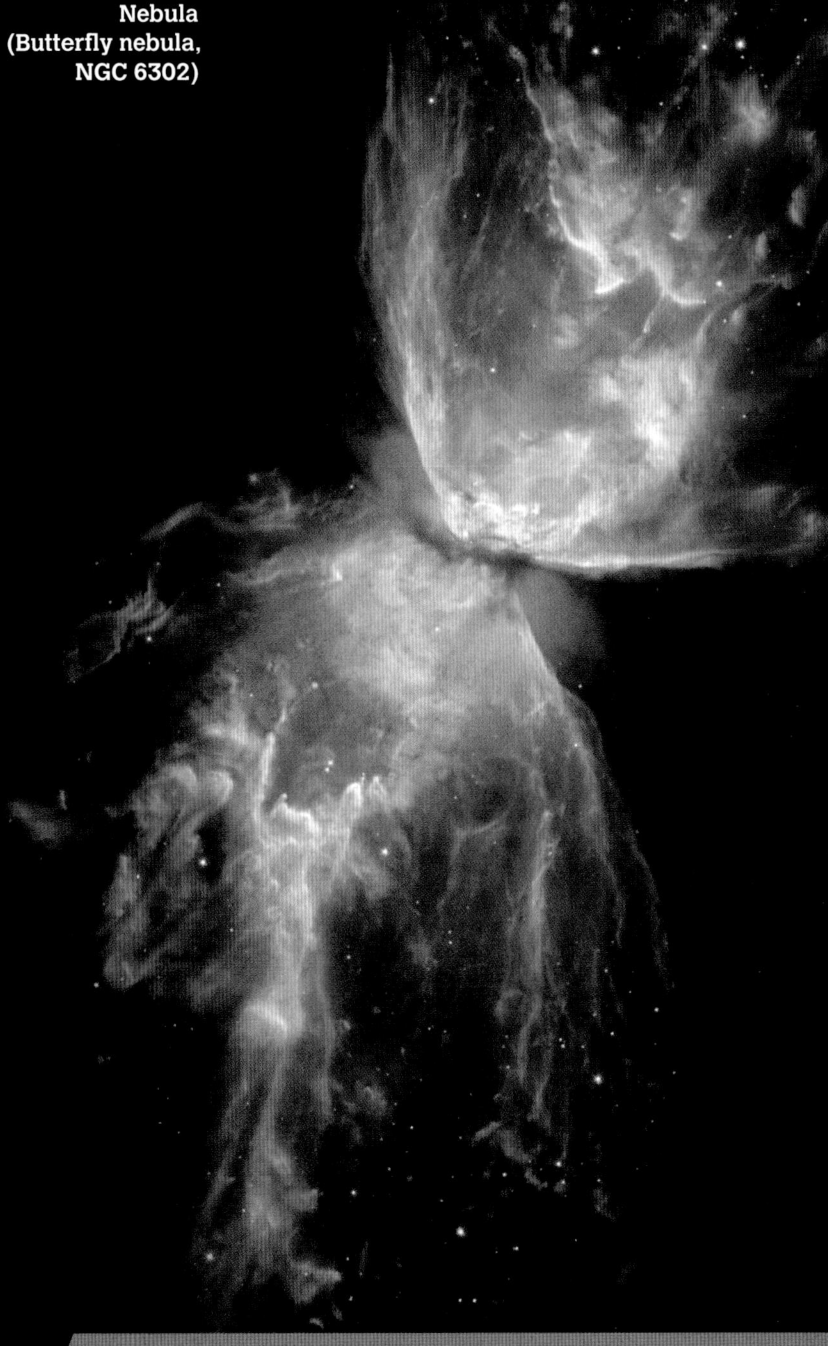
Nebula (Butterfly nebula, NGC 6302)

INDEX

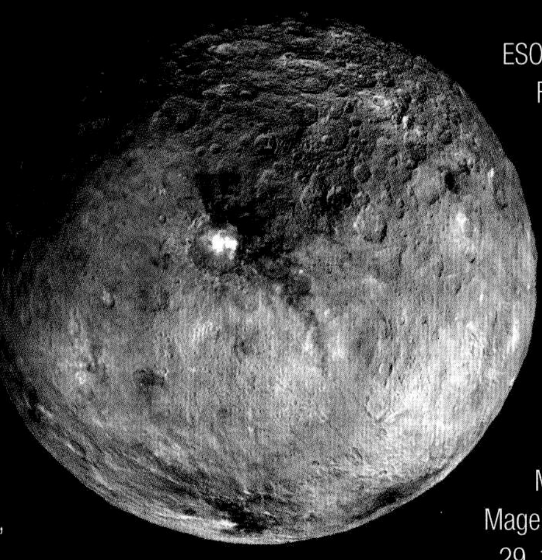

Dwarf planet (Ceres)

Constellation (Hydra)

Moon (Europa)

INDEX

Ra
(Egyptian sun god)

R

S

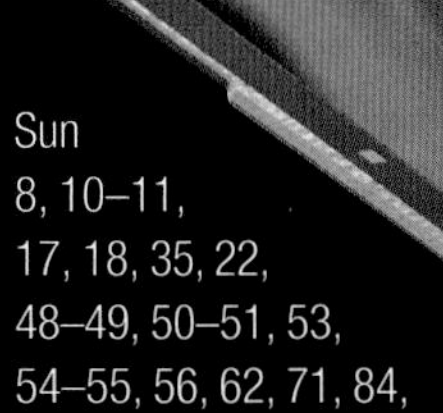

Space telescope (Spitzer)

T

U

V

WXYZ

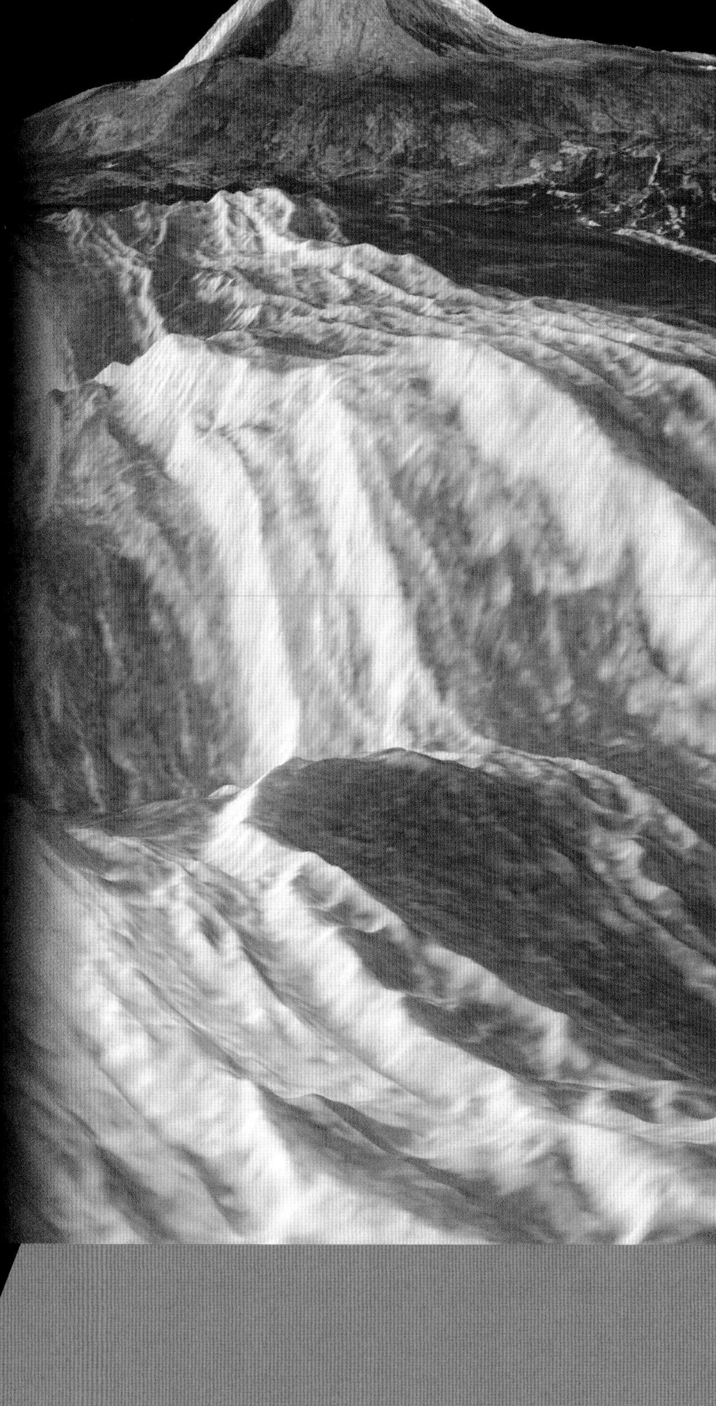

Volcano (Maat mons)

INDEX

Photos ©: 123RF: 117 top right (Andrew F. Kazmierski), 145 bottom center right (digidreamgrafix); Alamy Images: 69 top right, 195 bottom right (AF archive), 142 right-143 left (Chris Howes/Wild Places Photography), 192 top right (Chronicle), cover spine helmet, back cover top right helmet, 123 top right (Dmitry Zimin), 160 top right (Everett Collection, Inc.), 149 bottom (Felix Lipov), 126 bottom (Florilegius), 122 bottom center (Jay Wanta), 13 top right (Johan Swanepoel), 137 bottom right (John Gaffen), 90 (John White Photos), 193 background (Linus Platt), 67 top center (Michael Schmeling), cover bottom right pod, 123 bottom right (Michelle Enfield), 75 bottom right (Mireille Vautier), 105 center bottom right (Moviestore collection Ltd), 71 center right, 125 bottom center (Paul Fearn), 21 center (Pictorial Press Ltd), 122 bottom left (RGB Ventures/SuperStock), 35 bottom center left (Science History Images), 67 bottom center (Seaphotoart), 147 bottom left (SIPA Asia/ZUMA Wire), 21 center left (SOTK2011), 20 center right (The Print Collector), 21 bottom (Tibor Agocs), 139 top right (Xinhua); AP Images: 131 center top (ITAR-TASS), 85 center bottom (Tu meifei - Imaginechina), 143 center right top (Zhu lan - Imaginechina), 131 center right; Bonhams: 136 right, 137 top right; Courtesy of Caltech: 16 center bottom right, 22 center left, 45 center bottom left, 45 center bottom right, 55 top right, 116 bottom right; Dr. Margaret Geller: 182-183 left, 183 top right, 183 center right top, 183 bottom right, 183 center; Dreamstime: 117 top center (Crazya88), 35 center top far left (Frogtravel), 123 top left (Konstantin Shaklein), 131 tortoise (Maxim Petrichuk); Flickr/NASA: 149 center, 158 center top, 161 top right, 161 bottom right, 189 top left, 189 bottom right, 195 center left; Fotolia: 109 center left (Archivist), 19 top left (Jean-Jacques Cordier), 16 center top right (nickolae), 51 bottom left, 206 left (Paolo Gallo); Getty Images: 131 bottom center right (Bettmann), 91 bottom right (DEA/G. DAGLI ORTI), 122 top right (DIETER NAGL/AFP), 35 center top right (Fine Art Images/Heritage Images), 33 top right (Ronald Martinez/Stringer), 35 center top far right (Ronaldo Schemidt/AFP), 68 center top left (STR/ Stringer), 74 (ullstein bild Dtl.), 109 bottom right (UniversalImagesGroup); GMTO Corporation/Giant Magellan Telescope: 42-43 left; IAU/Sky & Telescope: 31 top center right; Institute of Space Systems, TU Braunschweig: 68 right-69; iStockphoto: 100 top (abadonian), 19 center right (advettr), 145 center far right (AndreaAstes), 131 bee (Antagain), 191 bottom right (Berezka_Klo), 18-19 background (bjdlzx), 70-71 moons (Delpixart), 164 top right (DieterMeyrl), 131 mouse (dra_schwartz), 19 center left (dzika_mrowka), 62 top left (Eerik), 124 bottom center (ekapol), 145 center left (Elen11), 174 top (Eloi_Omella), 125 top right (fcdb), 62 bottom left (Flory), 30 top center left (franksvalli), 102 top right (GeorgiosArt), 131 rat (GlobalP), back cover top center sketch (ivan-96), 17 top righ, 198 left (jodiecoston), 145 bottom center left (Joel Carillet), 28 top left (Jonlynch), 68 center top right (julichka), 19 bottom right (karandaev), 18 bottom left (La_Corivo), 123 bottom center left (Leadinglights), 164 top left (macida), 62 center (manjik), 19 center (MattStansfield), 123 center top, 124 bottom center background, 124 bottom center (Meinzahn), 75 center (narxx), 32-33 bottom background (Nikolas_jkd), 18 top left (parameter), 142 bottom left (pipcoalan), 85 bottom center (powerofforever), 131 fruit flies (rob_lan), 67 bottom left (sara_winter), 71 top right, 75 center left (scotspencer), 171 top right (sedmak), 18 bottom right, 30 top left, 131 top left (shaunl), 125 top backgrounds (Siberian Photographer), 131 bullfrog (tacojim), 145 bottom right (taka4332), 131 guinea pig (tap10), 145 center (Tomwang112), 50 bottom left (vladimirts), 56 center, 102 center (Whiteway), 145 bottom left (Xsandra), 88 top right, 102 left, 103 bottom center (ZU_09); Ken Crawford/www.imagingdeepsky.com: 174-175; LBTO: 89 top right; Library of Congress: 17 center bottom right (Cleveland, O.: W.J. Morgan & Lith., c1889 April 8), 127 left (Jet Lowe), 188 center left (Londini, Jussu Societatis Regiæ ac Typis Josephi Streater. Prostat apud plures Bibliopolas. Anno 1687), 35 center bottom right (Orren Jack Turner), 16 center top far left, 31 top center left, 204 right (Sidney Hall), 16 bottom left, 35 top left, 35 center top left; Lowell Observatory: 114 top left; Max Planck Institute for Radio Astronomy, Bonn, Germany/ Peter Sondermann, VisKom/City-Luftbilder: 23 bottom; National Science Foundation/Josh Landis: 109 bottom left; NOAA: 16 center top left, 89 bottom, 103 bottom left; NRAO/AUI/NSF, I. de Pater: 16 center bottom left; Rex USA/USA TODAY Network/Sipa USA/Shutterstock: 194 center right; Science Photo Library: 23 center bottom (MAX-PLANCK-INSTITUTE FOR RADIO ASTRONOMY), 183 center right bottom (Volker Springel/Max Planck Institute for Astrophysics); Science Source: 14-15, 24-25, 28 bottom left, 131 center far right, 168 right (Babak Tafreshi), 165 top right (David Parker), 68 center top center, 146 top right (Detlev van Ravenswaay), 34 (Getty Research Institute), 20 center left (Gianni Tortoli), 17 center top center, 32 bottom right (John Chumack), 22 bottom right (John R. Foster), 32-33 top background (Juan Carlos Casado/StarryEarth), 19 bottom left (Larry Landolfi), back cover top right meteorite, 77 bottom left (Manfred Kage), cover center right gloves (Mark Williamson), 44, 103 top, 124 bottom right, 133 center far left, 146 center (NASA), 11 top center, 16 bottom center, 20 bottom, 131 bottom right (New York Public Library), 193 top left (NOAA/Nature Source), 132 top right (Novosti Photo Library), 21 top left (NRAO/AUI/NSF), 130 (RIA Novosti), 92 bottom center (Robert and Jean Pollock), 22-23 starry background (Robert Gendler/Stocktrek Images), 91 Holmes (Rolf Geissinger/Stocktrek Images), 76 bottom (Royal Astronomical Society), 188 bottom (Sanford Roth), 98 top right (Smithsonian Institution Libraries), 45 center top left (Spencer Sutton), 193 center left (SPL), 20 top left, 20 center, 56 top right, 98 top left, 101 top center, 126 top; Shutterstock: 131 top right (3Dstock), 145 center far left (Akira3288), 19 bottom center (EpicStockMedia), 77 top left (itechno), 145 center right (Ivan Smuk), 16-17 star map (Marzolino), 26-27 star map, 28 right-29left, 28-29 star map (shooarts), 192 bottom right (Solent News Photo Agency/ REX), 70 bottom right (Yaroslaff); Space Telescope Science Institute: 185 center; UN Photo/Yutaka Nagata: 161 center bottom left; University of Arizona/NASA/JPL: 106-107; University of Hawai'i: 195 bottom left; W.M. Keck Observatory/Imke de Pater (UC Berkeley): 105 center top left; Wellcome Images: 35 bottom far left (Justus Sustermans), 101 top left (W. Ward), 31 top left, 35 center bottom left, 203 left; Wikimedia: 17 center bottom left (Blueshade), 84 Europa (Database of Asteroid Models from Inversion Techniques), 28 bottom center (ESA/Hubble & NASA), 116 bottom left (Hugo van Gelderen/Anefo), 116 bottom center left (Joop van Bilsen/Anefo), 27 top center (Mikulski Archive for Space Telescopes (MAST), STScI, and NASA), 33 top left, 40 bottom right, 43 top right, 43 center right bottom, 88 bottom right, 97 Telesto, 110 center left, 110 bottom right, 115 bottom right, 125 top left, 131 center left, 133 top center left, 133 top left (NASA), 73 center right bottom (NASA/GSFC/ Arizona State University), 79 top right (NASA/JPL-Caltech/Arizona State University), 43 bottom right (NASA/MSFC/David Higginbotham/Emmett Given), 165 top center (-r.c. (talk)), 127 right (SpaceX), 146 top center (Viktor Patsayev).

All other images © NASA, ESA, and ESO Science Outreach Network.

The publisher would like to give particular thanks to the following people for their help: Dawn Bates, Ali Scrivens, John Goldsmid, Rachel Phillipson, Laura Buller, Terry Buck, Marybeth Kavenagh, Debbie Kurosz, and Ed Kasche; Dr. Carsten Wiedemann of Braunschweig University of Technology, Germany; Dr. Margaret J. Geller, Harvard-Smithsonian Center for Astrophysics; Dr. Rosaly Lopes, NASA's Jet Propulsion Laboratory; Professor Paul Hickson; Jerry L. Ross; Jefferson Hall, NASA's Jet Propulsion Laboratory.

ACKNOWLEDGMENTS